I0889623

Translator Self-Training Spanish

Practical Course in Technical Translation

Morry Sofer

Second Revised Edition

Schreiber Publishing
Rockville, Maryland

Second Revised Edition

Acknowledgment: Section 2, Assignment 4, and Section 3, Assignment 4, are used with the kind permission of the World Bank.

Library of Congress Control Number: 2005924019
ISBN 0-88400-307-8

Printed in the U.S.A. by Schreiber Publishing, PO Box 4193, Rockville, Maryland 20849
www.schreiberpublishing.com

Preface

This practical course is for anyone who would like to pursue technical translation in Spanish and English, in such areas as legal, medical, financial or other kinds of non-literary documents (literary translation is a different discipline from technical).

The course is based on the practical experience of a major translation company which has been translating—routinely and successfully—Spanish-into-English and English-into-Spanish documents in all technical areas for almost twenty-five years. The assignments in this book are taken from real-life documents which were translated by this company, with names and other information changed in some cases to protect proprietary information.

This book grew out of a correspondence course offered about fifteen years ago, which became a three-ring binder program for translator's self-training ten years ago. During those ten years, hundreds of translators have used this program each year, and many have offered unsolicited praise for this program, letting us know that it has improved their translation skills, and in more than a few cases helped the user pass accreditation or certification tests for such organizations as the American Translators Association (ATA).

Nothing sharpens translation skills more than the actual act of translating. But one needs an entry point, either into the technical translation field in general, or into new areas of translation, if one has already been doing one kind of technical translation, such as legal. This course provides the entry point.

We at Schreiber Publishing take a personal interest in translators in all languages and subjects, and are always happy to hear from you. By the way, we also have an advanced program for Spanish medical translation and another one for Spanish legal translation, which you may want to consider once you complete the ensuing assignments.

Enjoy the course!

Morry Sofer

Table of Contents

Section 1. General

Introduction

Welcome to *Schreiber's* **Translator Self-Training Program**. We are very pleased you selected this program, which has been in use since 1992. It has helped aspiring translators get their start in this exciting field, while assisting practicing translators in improving their translation skills. Translation is one of the fastest growing professions in today's world, as different languages and cultures are coming closer together every day. Whether you wish to pursue translation full-time or as a sideline, you will find it to be a highly challenging and quite profitable occupation. The purpose of the following exercises is to help you improve your skills as a Spanish-into-English translator.

This program is designed to help you work at your own pace and gain experience through translation practice assignments. These assignments, excerpted from actual translation jobs, were selected to help you determine your own level of comfort and expertise, so that you may go on from there to choose the type of translation work you may like to pursue.

To derive the maximum benefit from these texts, you should carefully read the General Comments, the guidelines, and the specific notes which discuss each assignment. Every comment and every guideline represents years of experience in professional translation work. By carefully following this program, you will be able to achieve *in just a few days* what has taken others years to learn. You will be guided through many of the pitfalls of translation, taught how to avoid them, shown short-cuts in accomplishing a complex piece of translation work, and be given the benefit of comments designed to show you how to produce a translation acceptable to major clients such as government, industry, the legal world, and many more.

Before you read further, we strongly urge you to read Appendix A: Requisites for Professional Translators, and Appendix B: Translation Techniques.

The above-listed Appendices will give you a good overview of how translation work is to be approached and practiced, and how to position yourself in the translation market. Try to keep the contents of these chapters in mind as you translate the following assignments.

You are presented with four different texts, which, for most people, represent an increasing level of difficulty (although if you happen to be a lawyer, for example, it is possible that you will find the second assignment the easiest of the four). They are:

1. **General**
2. **Legal**
3. **Chemical/Medical**
4. **Business/Finance**

At a minimum, you may want to do the first and second assignments. Those two represent a vast body of material which is being translated in the United States on a regular basis, but which does not require a highly specialized vocabulary or technical background. The third assignment requires access to chemical and medical terminology, which can be managed by any intelligent translator. The fourth assignment should be approached with caution. Here one must have an understanding of the world of finance, which not all of us do. For your own self-assessment, you may want to try it (since we provide you with the translation, you can compare the results and see for yourself how you did). But if you find it too difficult, don't be discouraged. It is clearly not for everyone.

We hope you enjoy working with these texts, and we urge you to please be patient with yourself. One does not become an accomplished translator overnight. To learn, one must make mistakes. If your heart is in it, and if you love the challenge of working with two languages you happen to know well, then you are sure to get a great deal out of this program.

A Word about Spanish Translation

In the United States today there are large Hispanic populations hailing from different parts of the Spanish-speaking world, and whose Spanish dialect has a strong influence on the local Hispanic scene. Thus, for example, in Southern California there is a strong Mexican influence in the local Spanish; in Miami there is a strong Cuban influence; in parts of New York City a strong Puerto Rican influence, and so on. All of this impacts on translation work, both from and into Spanish. As a translator, what are you to do?

To use an example: A publisher in Illinois asked a translation company in Maryland to translate an English primer into Spanish. This textbook was targeting Hispanic audiences in different parts of the United States. It soon became clear to the Spanish translator that even simple, everyday words like "ball," do not have a uniform equivalent throughout the Spanish-speaking world. In South America the word is *pelota*. To many Puerto Ricans it is *bola*. One of the basic rules of translation is to choose the best equivalent possible and stick to it, rather than offer other alternatives. Here, however, it was necessary to offer more than one choice.

It can get even more complicated when we translate a text from Spain, because of the many cultural differences between the mother country of the Spanish language and the cultures of the New World. This writer was called upon by a major U.S. cultural institution to translate a book of artistic photography from Spain. It so happened that the author/photographer took those pictures in remote corners of Spain, where Spanish is mixed with local gypsy and other words which do not appear in any Spanish dictionary. It became necessary to get in touch with the author in Madrid, to get her explanations of those words.

In doing the assignments in this kit, keep those problems in mind and look for the best solution for each one of them. In translation, besides finding the right meaning of each word, you also need to find a way to present your rendition in a way that both preserves the overall meaning of the original text, and conveys it clearly and accurately to your reader.

General Comments

The following comments are offered with two purposes in mind: the immediate purpose is to help you with the assignments. The long-term purpose is to serve as reference material for actual translation work. Keep them for future use.

A reminder:
The "source language" is the language you translate from.
The "target language" is the language you translate into.

Every language has its own "personality," so to speak. When you translate, pretend you are two different persons - an "English Person" and a "Spanish Person." The Spanish Person reads the original Spanish text, and becomes familiar with it. The Spanish Person then turns to the English Person, and says to him/her, "**Okay, now it's your turn to say the same thing your way**." Now the English Person takes over, and starts writing the same text in English. You will be constantly switching back and forth between the Spanish Person and the English Person, as the first checks the Spanish word, phrase, sentence structure, and then the second thinks in English and comes up with the right equivalents. Here are some of the main language issues you will have to deal with:

Clarity

We each have our own personal style of expressing ourselves - in any language. Some people make their point clearly and succinctly, while others use more words than necessary, including esoteric words, to make their point. The same holds true in translation. You can translate the phrase, "Al pan, pan y al vino, vino," as "To call a spade a spade," or you could also translate it correctly as, "One should always speak forthrightly and call things by their proper name." In the first instance, only six words were used; in the second, twelve. Most of the time, the first type of translation is better, because it is simpler and more readily understandable. Make it your point to pursue simplicity whenever possible.

Consistency

Most of the time, if you start out translating the word "*casa*" as "house," stick to "house," and don't start suddenly shifting to "building," or vice versa. The reader may become confused, wondering why the word suddenly changed. Generally in English, building is bigger than house, and one may wonder whether the dimensions of that particular structure might have changed. If you are familiar with word-processing you know how easy it is to do a "global search and replace" throughout the text, so that if halfway into the document you decide that "building" is better than "house," then go back to the beginning and do a global change, replacing the word "house" with "building."

Sentence Structure

It is difficult, yet very important, to ensure that sentence structure is correct and makes sense, yet still retains the flavor of the original text, particularly if this is written in a flowery, elaborate, or otherwise unique style. This fine line is achieved mainly through practice, but it is definitely helpful to read the original several times before even beginning the translation process in order to gain a good idea of the author's style.

It is not necessary to follow the structure of a sentence literally. A run-on sentence which may make sense in one language, may not make sense in the other, and therefore should be divided into two sentences (or more). The converse is also true. You should also avoid being too literal when it comes to word order, or your text will sound stilted, awkward, and artificial. You will need to rely on your own good writing skills in determining the most sensible word order and punctuation so that your translation flows smoothly.

Format

Format should be followed as closely as possible. Ideally, every page of translation should look exactly like the corresponding page of the original, except, of course, for the language. There is a very practical reason for this. It helps the reader, who may or may not know the source language, compare the two texts, make sure all the parts are there, and check a particular word or phrase in the source text. This, however, is not always possible, since a page of translation at times is longer than the same page in the source language. Nevertheless, every effort should be made to keep the format as similar as possible to the original.

Terminology

The use of the right terminology is critical in any translation, especially in any technical subject. Here the right kind of dictionaries and reference literature becomes very critical. Keep in mind, however, that dictionaries, no matter how good, are never complete, and do not provide all the answers. The other source of information is people who work in the field the particular translation covers. Their input is often more valuable than that of the dictionary, since they are the ones who work in the field and represent the potential readers of your translation.

The first thing to keep in mind about terminology - as was discussed before - is consistency. If you translate a word a certain way, do not change in the middle of the text to another synonym of the same word, even though it may be accurate. This will throw the reader off. The second thing is to make sure you are using the right terms, rather than trying to approximate them. Keep a list of those terms to maintain consistency throughout the document, and also for future use. If you continue to translate in the same technical field, these lists will become your most valuable translation tool.

Numbers

The use of commas and periods is reversed in numbers over 1000 (1.000 in Spanish becomes 1,000 in English), and in decimals (0,5 in Spanish, becomes 0.5 in English). This is one example of cultural differences which must be closely observed.

Punctuation

Quotation marks are appropriate in English, but this is not always so in Spanish. Semicolons have their place in both languages, but not necessarily in the same places. These are just a few examples of how even a highly accurate translation will not always mirror its original text. You will often be called upon to be the judge, which involves a combination of excellent writing skills, knowledge of rules of grammar and punctuation, plus common sense.

Errors in the Source Text

Sometimes you will find that the original has errors, such as a misspelling where it is clear by the context that the author intended to say "uno" and not "un," or the sequence in a series of numbered clauses is wrong, or a name that has been in upper case letters suddenly appears as lower case, for no apparent reason. You can do a number of things in situations such as these. You can go ahead and make the correction and leave it at that. You can translate the text as is, and use the term [sic] after the error, to alert the reader. Or you can make a translator's note on a separate sheet of paper. Ultimately, the client will have to decide what to do with it.

Untranslatable Words or Terms

You may run into a word, a term or a phrase which cannot be translated, or which should be left in the original language (such as, for instance, "Junta"). If there is a need to explain it, it can be done in a footnote or in parentheses (or brackets, which makes it clearer that the bracketed text is not part of the original) following the word (Junta [military ruling group]). As a general rule, however, it is best not to burden your translation with footnotes or parenthetic or bracketed text.

In Conclusion

The above notes are by no means exhaustive. They do not cover all the issues facing a translator. Language is not mathematics. Keep this in mind, and do the best translation job you possibly can.

Guidelines for Translating the Assignments

1. We are providing you with a Spanish text and an English translation of that same text. Pretend you don't have the English translation. If you read it before you do the assignment, you will not be able to do an honest job, and you will not derive the benefit of learning from your own mistakes.

2. Read the Spanish text once for general comprehension. Underline any difficult words or passages.

3. As you start the translation process by reading the text for the first time (see 2., above), keep track of time, so that when you finish you will know how long it took you to translate the text. This is very critical, because you want to increase your translation speed as you go along. The rule is simple: The more words you translate per hour, the more money you will make. A very low hourly word-count can earn you less than minimum wage, which is not at all desirable. (Note: The hours indicated at the top right hand corner for each assignment include two numbers. The first is for beginners while the second number is for an experienced translator. Therefore, an assignment that would take a beginner 5-7 hours to complete might take an experienced translator 2 hours. The timeframe includes research, draft, and final copy.)

4. Read it again. This time make a list of difficult words, and make any notes regarding things you need to clarify about the text (e.g., look up something in an encyclopedia to better understand some concept or piece of information).

5. Try to determine which dictionaries and reference books you need for this assignment, get hold of them, and keep them nearby.

6. Now you are ready to start translating. Preferably, you will be using a computer. If this is too intimidating, write it out the first time. It will take longer, but you will have more time to think. Remember, your objective is to translate on computer, not in longhand. These days, if you want to earn money translating, you have to submit text on disk. There are hardly any exceptions to this requirement.

7. Do a draft translation. Go over the draft. Do the best you can to make the English sound like good, clear, correct English, which does not sound like Spanish, and does not retain any Spanish forms in either style, format, or any technical points. Be sure you did not omit anything. Ask yourself if you succeeded in conveying the correct sense of every idea and every word in the source text. Be sure you are comfortable with every single word you chose in English, and that you have no doubts or hesitations. If, after all this, you do have

some questions, do not worry. No one expects you to do perfect work the first time around. You will do better next time.

8. Now that you have accomplished your translation, figure out the time it took you to reach your final version (including all the previous steps, namely first reading, research, draft, and final copy).

9. Make a list of the words, passages, and concepts in the text where you encountered difficulty, and keep for future reference. This kind of record keeping will help you improve your performance, and serve as a reminder not to repeat the same mistake twice.

10. Print out your corrected and edited text.

Guides for Self-Evaluation

1. Turn to the English translation in Section 3, and compare it to yours. Next read the notes concerning the pitfalls and salient points of this text very carefully, and see how they compare to your own experience with the text. They may not necessarily coincide with everything you experienced personally, but they do provide you with some tips on the do's and don'ts of technical translation.

2. The most important thing for you to keep in mind as you compare your translation to the one in Section 3, is that there is no one absolutely correct way of translating a text. Many words can be translated in more than one way. Many sentences can be structured in more than one way. Many ideas can be conveyed in different words. In fact, your translation may turn out to be just as good, and in some spots even better than the one in the book.

3. Keep track of the shortcomings of your translation. The areas you are interested in are the following:

   Omissions - did you fail to translate any particular word or phrase, or even paragraph?

   Format - does your format follow the original (breaking into paragraphs, for instance)?

   Mistranslations - did you mistranslate any particular word?

   Unknown words - were there any words you were not able to translate?

   Meaning - did you miss the meaning of any particular phrase or sentence?

   Spelling - did you misspell a word?

   Grammar - did you make any grammatical mistakes?

   Punctuation - did you mispunctuate or miss any punctuation marks?

   Clarity - did you fail to clearly convey the meaning of any particular part of the text?

   Consistency - did you call something by one name, and then by another without good reason?

   "Sound-alike" words - were you "tripped" by a word in the source text, whereby you mistranslated it because it looked the same, though it changes meaning in the target language (for instance, the word "actual" exists in both English and Spanish, but means two different things)?

   Style - are you satisfied with the way your translation reflects the style of the original (for example,the original is written in a clear, direct style, while the translation sounds more complex and indirect)?

These are some of the questions you have to ask yourself. As you go on to translate other texts, always keep these questions in mind as you self-assess your translation. By improving your performance in the above areas, you are on your way to becoming an accomplished translator.

Part One

Translation: Spanish into English

Section 2. Assignments

Assignment 1: General

5-7/2 hours

HUICHOLES. — Emparentados con los coras y descendientes de los bravos caxcanes o cascanes, los huicholes son una de las comunidades étnicas más representativas del Occidente. Las localidades donde se ubican pertenecen a la zona colindante de Jalisco y Nayarit, surcada por sistemas de serranías accidentadas y bellas. Allí, en sitios prácticamente aislados, estos indígenas continúan viviendo según sus antiguas tradiciones, las cuales regulan su existencia desde la cuna hasta la sepultura. Quizás el aspecto más evidente y singular de los huicholes se encuentra en sus vistosos trajes, confeccionados en manta blanca común, ornamentados con intrincados diseños multicolores, debido a la habilidad de las bordadoras, quienes con paciencia y estambres de colores van realizando los motivos consagrados por la tradición: estilizados venados, serpientes, liebres, pájaros; extrañas flores de pétalos simétricos; elegantes motivos geométricos representantes del sol, el agua, los montes. La fuente de inspiración para las consumadas obras de arte así fabricadas —como también las elaboradas a base de chaquira, pintura, y papel amate— se encuentran en las místicas visiones provocadas por el peyote, un pequeño cacto que crece espontáneamente en el estado de San Luis Potosí y posee efectos alucinógenos. Hasta ese estado los huicholes peregrinan anualmente y una vez en regreso, organizan unas ceremonias en el curso de las cuales consumen el peyote, como medio para acercarse a sus dioses y comunicarles sus peticiones. El peyote, conocido por los huicholes con el nombre de "hículi hualula", produce entre otras cosas unas maravillosas visiones que presentan espléndidos colores. Las sabias manos de estos artesanos las transforman en pequeñas obras de arte, tanto en prendas de vestir como en objetos rituales, entre los más conocidos de los cuales están los llamados "Ojos de Dios", cruces de caña entretejidas con estambres de vivo colorido. Actualmente los huicholes constituyen un pueblo pacífico, entregados a sus tradiciones y labores manuales, pero antaño fueron temibles guerreros, por mucho tiempo imposibles de conquistar. Hoy en día, estas comunidades de incomparables artistas están en vías de integrarse al progreso buscando un futuro más próspero.

EL MARIACHI. — Mundialmente famosos son los grupos musicales conocidos como "mariachis"; en todas partes su atuendo y su música llenan de alegría a propios y extraños, quienes de inmediato los identifican como símbolo popular de México. El mariachi es originario de Jalisco, uno de los muchos frutos del mestizaje cultural que caracterizan nuestra peculiar idiosincrasia. En principio se trataba de conjuntos integrados por instrumentos de cuerdas: guitarrón (propio de México), guitarra, dos violines y una vihuela. Cuando empezaron a dar serenatas y mañanitas, es algo que todavía no se ha podido establecer con absoluta certeza; pero es indudable su existencia ya en el siglo XVIII, aunque sin el nombre que ahora los identifica.

Es, según la leyenda, hasta la invasión francesa cuando el mariachi recibe este nombre, debido a la circunstancia de que los músicos solían tocar variadas melodías para amenizar las ceremonias y festejos de las bodas o matrimonios, mariage en francés, del cual por corrupción derivó la palabra mariachi.

Con el tiempo, a la ejecución puramente instrumental se añadió el canto de las melodías, y gradualmente se incorporaron instrumentos de aliento, como las trompetas, hasta lograr un efecto único en el mundo: esta peculiar combinación de voces humanas, cuerdas y metales, que distinguen el mariachi actual.

Si bien el mariachi es propio de todo Jalisco, los de Cocula pronto se consideraron los mejores y más famosos, no sólo por su técnica impecable sino por la auténtica creación de canciones especiales para esta modalidad musical; la composición presenta dificultades pues no es nada sencillo lograr la armonía de elementos tan dispares como los que conforman al mariachi; además el ritmo vivo y alegre de la música ranchera requiere una sincronización perfecta de todos los ejecutantes. El estilo musical propio del mariachi dio origen a esta música ranchera, que en el aspecto temático se nutre de la vena popular para producir coplas que narran y describen situaciones de la vida campirana o ensalzan las excelencias de pueblos y ciudades; pero el tema más importante es el amor, en todas sus formas.

Hacia el año de 1920 llegó a la ciudad de México, para establecerse en ella, el coculense Cirilo Marmolejo con su mariachi; en plena calle hacían sonar las vigorosas notas de sus alegres melodías, sosteniéndose con lo que la gente buenamente les daba al pasar. Tiempo después se instaló en Garibaldi, y en este lugar se incrementó su fama a tal grado que fue enviado como representante de México a la Feria Mundial de Chicago.

De aquellos tiempos a nuestros días, la popularidad del mariachi y de su bella música ha ido en constante aumento. Dondequiera que se escuchan sus sones característicos, el nombre de Jalisco se invoca con orgullo.

word count: 790

4-5/1.5 hours

Assignment 2: Legal

Doy fe de tener a la vista los documentos relacionados, de que la representación que se ejercita es suficiente de conformidad con la ley y a mi juicio para el presente acto. Conozco al segundo de los comparecientes y la primera por no serlo con anterioridad se identifica con la cédula de vecindad número de orden II guión ocho y de registro cincuenta y dos mil doscientos treinta y cuatro extendida por la municipalidad de Salta, municipio del Departamento de Belgrano; y quienes asegurándome hallarse en el libre ejercicio de sus derechos civiles, me manifiestan que otorgan escritura de ADOPCION, contenida en las cláusulas siguientes: PRIMERA: Exponen los comparecientes que iniciaron diligencias de adopción, mediante el acta de requerimiento, que para el efecto autorizara en esta ciudad con fecha veinte y tres de febrero de mil novecientos ochenta y ocho el Infrascrito Notario, de conformidad con la ley de Tramitación Notarial de Asuntos de Jurisdicción Voluntaria, por la que la Licenciada CRISTINA ORTIZ, en representación de los señores SAMUEL HUNTER y GRACE ELLIS, manifiesta su deseo de adoptar a favor de sus representados como hijo propio al menor JOSE CHAVEZ, y quien es hijo de la señora CECILIA CHAVEZ. El Ministerio Público en su dictamen opinó favorablemente, previo el informe rendido por la Trabajadora Social del Juzgado Quinto de Familia, designada para el efecto, manifestando que se puede acceder a lo solicitado mandando se otorgara la escritura correspondiente a efecto de que se inscriba en el registro Civil de esta capital en donde corresponde. SEGUNDA: Me manifiestan los otorgantes que por este instrumento ratifican su voluntad expresada en el acta antes indicada, y por la cual el mandatario especial judicial adopta a favor de sus representados los señores SAMUEL HUNTER y GRACE ELLIS, al menor JOSE CHAVEZ, como hijo propio, manifestando que por no contar con bienes dicho menor no se faccionó el inventario respectivo, por lo que los comparecientes cada uno con el carácter con que comparecen manifiestan su voluntad de que se otorgue este instrumento a efecto de que quede formalizada la adopción, solicitando al Registrador Civil de esta capital, proceda a anotar la partida de nacimiento bajo el Número DOS MIL TRESCIENTOS CUARENTA (2.340), folio CIENTO CUATRO (104), del libro CIENTO CINCUENTA (150), de nacimientos. TERCERA: Exponen los otorgantes, que están enterados de los derechos y obligaciones que contraen en este instrumento, y la facultad que tiene el menor adoptado de usar los apellidos de los adoptantes los señores SAMUEL HUNTER y GRACE ELLIS y de tenerse como hijo propio y representarlo legalmente cuando sea necesario. CUARTA: La señora CECILIA CHAVEZ, en los términos relacionados, acepta la adopción que por medio de su mandatario especial judicial hacen de su hijo menor indicado los señores SAMUEL HUNTER y GRACE ELLIS, dando los otorgantes cada uno con el carácter con que comparecen su conformidad con todas y cada una de las cláusulas de este instrumento. Yo el Notario doy fe de todo lo expuesto, de haber tenido a la vista el expediente de Tramitación Notarial de Adopción ya indicado, así como el dictamen del Ministerio Público antes relacionado, y que por designación de los otorgantes doy integra lectura a lo escrito e impuestos de su contenido, objeto, validez y efectos legales, de la obligación al Registro de Testimonio de esta escritura, la aceptan, ratifican y firman con el Notario que autoriza.

word count: 562

Chemistry

Compuesto de intercalación

Compuesto que consta de una red cristalina que actúa como un donante de electrones y átomos receptores de electrones "extraños" intercalados o difundidos entre los planos de la red. Un importante grupo de compuestos intercalados son los de grafito, donde el bromo, por ejemplo, puede actuar como receptor de electrones. El grafito es particularmente susceptible a este fenómeno debido a sus capas de cristales ordenadamente apiladas. Los nitratos metálicos anhidro tales como los nitratos de cobre y zinc también forman compuestos intercalados con el grafito. Otro ejemplo es el nitruro de trilitio, cuya estructura se compone de una serie de capas de nitruro de dilitio, entre las cuales hay una capa de átomos de litio. Esto incrementa marcadamente la conductividad de forma que el material se convierte en un electrólito sólido efectivo en baterías. Otras sustancias que tienen esta propiedad son el disulfuro de titanio y algunos dióxidos metálicos. El fenómeno no altera la estructura cristalina y es reversible. Los compuestos intercalados se usan para superconductores, lubricantes sintéticos, catalizadores y acumuladores. También se usan en investigación bioquímica; un compuesto basado en la acridina que puede intercalarse entre pares de bases de una hélice de DNA se emplea en la investigación del cáncer.

Alergia y sensibilización

Aumentan en las empresas la incidencia del asma, la sensibilización química múltiple, y las enfermedades epidérmicas. La definición clásica de sensibilización es que constituye una respuesta iniciada por el sistema inmunológico. Hay dificultades par definir "atopía". Debe definirse un sensibilizador como aquella sustancia que puede provocar una reacción alérgica en el individuo medio, o solamente en ciertas personas?

Monitoreo biológico

La producción o inhibición de un compuesto metabólico especial o proteína, o el daño cumulativo al DNA, pueden constituir un indicador sensible del grado de exposición a una sustancia tóxica. Además, el examen genético es cada vez más sofisticado, barato, y accesible al médico común. Sin embargo, plantea cuestiones éticas y científicas.

Cáncer y carcinógenos

Un problema que adquiere prominencia y provoca incertidumbre es de como identificar y clasificar los carcinógenos. Por ejemplo, recientes cambios introducidos al sistema de clasificación ACOIH establecen que si se trata de un carcinógeno animal y la epidemiología no ha aportado evidencia de que existe en los seres humanos, entonces es un A3. Pero esta conclusión es refutada, sin embargo, por la elección de sustancias a incluir en A3.

Productos químicos y toxicología

Se han presentado recientemente una serie de trabajos interesantes sobre toxicología, particularmente en relación a disolventes y metales. Hay asimismo una tendencia a prestar mayor atención a la inmunotoxicología (efectos sobre el sistema inmunológico) y la neurotoxicología (efectos sobre el sistema nervioso).

Lesiones musculoesqueléticas (MSI)

Las lesiones musculoesqueléticas constituyen actualmente la categoría principal de las enfermedades ocupacionales, habiendo aumentado su frecuencia en alrededor del 20% anual en los últimos tiempos. Al analizar las estadísticas sobre MSI, resulta evidente que las soluciones podrán encontrarse examinando los cambios que han tenido lugar en las empresas durante el período de incremento rápido.

Genética y reproducción

Existen tres áreas que provocan preocupación: los efectos genéticos y hormonales sobre la reproducción y los efectos sobre los hombres.

Genotoxinas - Es ampliamente aceptado el criterio de que ciertos productos químicos pueden provocar cambios genéticos, a su vez transmisibles a los hijos, a quienes ocasionan enfermedades. Sin embargo, la mayoría de la gente se sorprendería si supiera que, a pesar de años de investigación, no existe ninguna prueba científica absoluta de que consecuencias semejantes resulten de la exposición a algún producto químico.

Hormonas - Informes recientes han destacado la posibilidad de que el medio ambiente haya sido contaminado en gran escala por compuestos que interrumpen el funcionamiento del sistema endocrino. Algunos científicos creen en la actualidad que hemos inundado el medio ambiente con compuestos que contienen estrógenos.

Riesgos reproductivos en los hombres - La mayoría de los estudios acerca de las consecuencias adversas sobre el sistema reproductivo (defectos de nacimiento, abortos involuntarios, peso bajo al nacer) se han centrado en la exposición de la mujer a diversos riesgos. Lo que se estudia con menor frecuencia es el efecto sobre la reproducción de la exposición a riesgos sufrida por el padre, o la incidencia de su estilo de vida.

Fibras minerales

La fibra de vidrio u otras fibras minerales sintéticas son los principales sustitutos del asbesto. La industria del asbesto con insistencia ha destacado que dichos sustitutos no están exentos de riesgos. Ciertas enfermedades pulmonares y variedades de cáncer similares a las ocasionadas por el asbesto han sido observadas en operarios que manejan otras fibras; la evidencia sugiere que la mayoría de ellas son más seguras que el asbesto.

Campos electromagnéticos y otras radiaciones

Todos estamos continuamente rodeados de campos magnéticos, tanto en la empresa como en el hogar. El resultado de varios estudios, aunque de cierto modo no concluyentes y contradictorios, tiende a demostrar que existe un nexo entre la exposición a campos electromagnéticos y ciertos tipos de cáncer (en especial, tumores en el sistema nervioso central y leucemia infantil).

word count: 838

PRESTAMOS
y
COOPERACION TECNICA

TERMINOS Y CONDICIONES

El método para calcular la tasa de interés variable del Banco, que fue adoptado en diciembre de 1989 y se basa en una canasta de empréstitos, ajusta uniformemente la tasa de interés aplicable a los saldos pendientes de todos los préstamos aprobados desde el 1 de enero de 1990. De esa manera, y en consonancia con el carácter cooperativo de la institución, todos los prestatarios comparten por igual el costo y los beneficios de las variaciones en la tasa de interés. La tasa de interés que se obtiene con el método de fijar la tasa en el momento del desembolso seguirá aplicándose a los desembolsos de los préstamos aprobados antes del 1 de enero de 1990. Sin embargo, ambas tasas tienen el mismo margen sobre el costo de los empréstitos, que en 1994 fue del 0,5%.

Los nuevos empréstitos se destinan al financiamiento de desembolsos de préstamos aprobados con cualquiera de los dos métodos de determinación de las tasas de interés. El costo de los empréstitos para créditos con tasa fija en el momento del desembolso es el costo promedio ponderado de los nuevos empréstitos contraídos en los 12 meses anteriores, asignados al financiamiento de préstamos aprobados antes del 1 de enero de 1990. El costo de los empréstitos para créditos con tasa de interés variable es el costo promedio ponderado, durante los 6 meses previos, de todos los empréstitos pendientes asignados a la canasta que financia los préstamos autorizados desde el 1 de enero de 1990.

El tipo de interés para los nuevos desembolsos de préstamos con tasa fija en el momento del desembolso fue del 6,14% durante el primer semestre de 1994 y del 5,63% durante el segundo. El interés aplicable a los saldos pendientes de préstamos con tasa variable fue del 6,95% durante los primeros seis meses de 1994 y del 6,81% durante el segundo semestre. Los plazos de vencimiento para los préstamos concedidos con fondos del capital ordinario durante 1994 fluctuaron entre 20 y 30 años.

Durante 1994 se concedieron seis préstamos de la nueva ventanilla en dólares de Estados Unidos del capital ordinario. Estos préstamos se denominan y desembolsan únicamente en esa moneda, con plazos de vencimiento de 10 años como máximo y la posibilidad de optar por una tasa fija o una tasa flotante vinculada a la LIBOR y basada en el costo para el Banco del financiamiento de ese tipo en dólares de Estados Unidos, con el mismo margen sobre el costo de los empréstitos que los demás préstamos del capital ordinario. El propósito de este programa es estimular la actividad del sector privado proporcionando financiamiento en condiciones más compatibles con las necesidades de los prestatarios del sector privado que recibirán préstamos.

En 1994 se concedió un préstamo del capital ordinario en moneda del país prestatario, con un interés del 4% y un plazo de vencimiento de 20 años.

Las tasas de interés aplicables a los préstamos en moneda local con cargo al Fondo para Pequeñas Empresas (FPE) oscilaron entre el 1 y el 4%, según el grado de desarrollo del país y la índole del proyecto, con períodos de gracia de 5 a 10 años y plazos de vencimiento de 25 a 40 años. La tasa de interés sobre los préstamos en moneda convertible con cargo al FOE que se aplicó a los países menos desarrollados fue del 1% durante los primeros 10 años y del 2% de ahí en adelante, con períodos de gracia de 10 años y plazos de amortización de 40 años.

Además del interés sobre los préstamos, el Banco aplica una comisión crediticia del 0,75% para los préstamos provenientes del capital ordinario aprobados después del 1 de enero de 1989, del 1,25% para los préstamos del capital ordinario aprobados antes de esa fecha y del 0,5% para los provenientes del FPE, así como una comisión única del 1% para inspección y supervisión que se aplica a todos los préstamos, aunque durante el primer semestre otorgó una exención de 75 centésimos de punto porcentual de dicha comisión.

Los préstamos que no se otorgan directamente a los gobiernos de los países miembros o a instituciones intergubernamentales de desarrollo llevan garantía de los propios gobiernos o de organismos de gobierno. Los financiamientos de pequeñas empresas, que no requieren garantía, se extienden en moneda local can plazos de hasta 40 años y períodos de gracia de hasta 10 años. A estos financiamientos se les aplica una comisión del 1% anual. La cooperación técnica se otorga con carácter reembolsable, no reembolsable o de recuperación contingente. La de recuperación contingente, que representa una porción reducida del total, debe amortizarse únicamente si, como resultado de la operación, el Banco u otra institución de financiamiento externo concede posteriormente un préstamo para la ejecución de un proyecto o programa.

word count: 755

Section 3. Translation of Assignments and Notes

Assignment 1: Translation

THE HUICHOLES. - Related to the Coras and descendants of the brave Caxcans or Cascans, the Huicholes are one of the most ethnically representative communities of the West. They are located in the contiguous zone of Jalisco and Nayarit, scored by rough yet beautiful mountain ranges. There, in practically isolated spots, these indigenous peoples live according to their ancient traditions which govern their existence from the cradle to the grave. Perhaps the most telling and unique aspect of the Huicholes is their showy costumes, made up from ordinary white fabric and ornamented with intricate multicolored designs thanks to the skill of the embroiderers, who, with patience and colored worsted yarn, create the motifs consecrated by tradition: stylized deer, serpents, hare, birds; strange flowers with symmetrical petals; elegant geometric motifs representing sun, water, and mountains. The source of inspiration for the accomplished works of art thus created - as with those created on the basis of embroidery beads, paint, and ficus paper - is found in the mystical visions produced by peyote, a small cactus which grows naturally in the state of San Luis Potosí and which has hallucinogenic effects. The Huicholes travel to this state annually, and upon their return put on ceremonies during which the peyote is consumed as a means of getting closer to their gods and to communicate their wishes to them. The peyote, known to the Huicholes as "hículi hualala," causes, among other things, marvelous splendidly colored visions. The skilled hands of these artisans transform them into small works of art, from articles of clothing to ritual objects, among which the most well known are those called "God's eyes," crosses made from reeds interwoven with vividly colored worsted yarn. The Huicholes today are a peaceful people, dedicated to their traditions and to manual labor, but once upon a time they were fearsome warriors, unconquerable for a long time. Today, these communities of incomparable artists are in the process of succumbing to progress in search of a more prosperous future.

THE MARIACHIS. - The musical groups known as "Mariachis" are world famous; their ceremonious apparel and their music fill both natives and strangers everywhere with joy, and make them readily recognizable as a popular symbol of Mexico. The Mariachis originated in Jalisco, one of the many products of cultural blending which characterize our particular idiosyncrasy. At first, they were groups made up of string instruments: a "guitarrón" (large guitar typical of Mexico), a guitar, two violins, and a vihuela (a special kind of guitar commonly played in tropical areas). When they began their evening and early morning serenades has not been established with absolute certainty, but undoubtedly they were already in existence by the eighteenth century, although without the name by which they are known today.

According to legend, it was at the time of the French invasion that the Mariachis received their name, due to circumstances whereby the musicians would play various melodies for entertainment at ceremonies and celebrations of weddings or marriages, *mariage* in French, the corrupt form of which gave rise to the term Mariachi.

In time, the singing of melodies was added to the purely instrumental execution, and gradually, wind instruments, such as the trumpet, were also incorporated, to produce an effect unique in the world: that particular combination of human voice, string and brass which distinguish today's Mariachis.

Although the Mariachis are typical of all of Jalisco, those from Cocula were soon considered the best and most famous, not only for their impeccable technique, but also for the authentic creation of special songs for this kind of music. The composition presents difficulties, for it is no simple matter to produce harmony among such disparate elements as those which make up the Mariachi band; besides, the gay and joyful rhythm of the ranch music requires a perfect synchronization by all the performers. The musical style typical of the Mariachis originated this ranch music, which in terms of its theme is nourished by the popular vein to produce ballads which narrate and describe rural life situations or extol the superior qualities of villages and cities. The most important subject, however is love in all its forms.

Around 1920, Coculan Cirilio Marmolejo came to Mexico City to establish himself there with his Mariachi band. He would produce the sounds of the vigorous notes of his joyful melodies in the middle of the street, supporting himself with what he was given by the passersby. Later, he established himself in Garibaldi, and there his fame grew to such an extent that he was sent to the World's Fair in Chicago to represent Mexico.

From those times through today, the popularity of the Mariachis and their beautiful music has been constantly growing. Wherever their characteristic sounds are heard, the name of Jalisco is intoned with pride.

Assignment 1: Notes

Occidente (line 2) - *West* is the appropriate translation for *occidente*. *Occidental*, though technically correct, is not a geographic term, so to speak. In Spanish, the west is referred to as *el Occidente* (hence it is capitalized) and the east is *el Oriente*.

zona colindante (line 3) - Here one must read the text carefully, since it could be easily taken to mean "the zone adjacent to Jalisco and Nayarit," when, in effect, Jalisco and Nayarit are contiguous, and the Huicholes actually live in those two states.

serranías accidentadas y bellas (line 4) - Rather than translate this as rough and beautiful mountain ranges which does not quite make sense, we may take the liberty of changing *y* to *yet*, which is in keeping with the author's meaning and sounds better in English.

sepultura (line 6) - The actual meaning of *sepultura* is *burial*, but although the author is using the word to connote life's end, we decided to use the equivalent of *cradle* (as opposed to birth), giving us *grave*. Here the meaning is preserved not by translating literally but rather by choosing the expression that sounds best to the English ear.

chaquira (line 12) - You may have to check a few sources of reference in order to find the meaning of *chaquira*, embroidery beads used for many purposes, including currency, decoration, and artwork. The same case occurs with *amate* (line 13) which is ficus or, in this instance, refers to the paper made from the ficus plant or tree. Note, however, that both these terms are actually very regional to this part of the world. The practice of using beads for everyday transactions, now incorporated in other activities, is an ancient custom in this area. The same occurs with *papel amate*, for since the ficus is predominant there, it has been the source of paper for native tribes for centuries.

Hasta ese estado los huicholes peregrinan ... (line 15) - This is an example of changing the order within the sentence. The Spanish version is a more poetic version of the translated form of the phrase, but is nonetheless correct. The risk in trying to stay closer to the more intricate way of writing it is that it might end up sounding like an overly literal version of the original.

sabias manos (line 19) - *Wise*, *smart*, or *knowing* would be technically correct in this case, but are too literal. These possible translations lead to *skilled*, which is correct, as well as in keeping with the author's particular descriptive style. (*Wise*, however, would have been the next best choice.)

frutos (line 29) - The author uses the word *frutos* in terms of results or *products*, which we chose as the most exact translation, particularly since it is followed by *mestizaje cultural* (line 29), which refers to a process, namely that of cultural blending.

mañanitas (line 32) - Though these are defined in the dictionary as serenades, we also need a word for the preceding *serenatas*, so we combined them both, but still distinguished between the two types by establishing the time of day at which they were played.

metales (page 2, line 7) - The distinction among instruments refers to string versus *brass*, not metals, as you may have put instead.

; (line 11) - It made more sense to split this sentence into two parts so it would not be a run-on (especially with the use of a second semicolon in the same sentence in line 11, page 2). Often in Spanish you will encounter run-on sentences, especially where the style is poetic or otherwise complex (such as in legal documents), but there is certainly no need to follow such a structure blindly, particularly if the result in English will be nothing more than confusing. Further on, the third time a semicolon appears in that paragraph (line 11, page 2), it is actually unnecessary because the clauses are not independent of one another, so a comma is more appropriate.

coplas (line 16) - The dictionary gives several words for *coplas*, but *ballads* fits best from the descriptions given (even if you know nothing about Mariachis) of the traveling musicians, who sing of love.

I attest that the relevant documents are before me, that the representation here exercised sufficiently conforms with the law and with my judgment for the present proceeding. I know the second of the appearing parties, and not knowing the first beforehand she is identified by identity card serial number II hyphen eight and register fifty two thousand two hundred and thirty four, issued by the municipality of Salta, Department of Belgrano; and having assured me that they are freely exercising their civil rights, declare that they grant the deed of ADOPTION, detailed in the following clauses: FIRST: The executing parties state that they initiated the adoption proceedings by the requisition document, which authorizes that in this city, on the twenty-third of February, one thousand nine hundred and eighty eight, the Undersigned Notary, in accord with the Law of Notarial Proceedings of the Voluntary Jurisdiction, by which CRISTINA ORTIZ, Esquire, representing Mr. SAMUEL HUNTER and Ms. GRACE ELLIS, expresses her desire to adopt, on behalf of the parties she represents, the minor JOSE CHAVEZ, who is son of Ms. CECILIA CHAVEZ, as their own child. The Attorney General's Office gave its opinion in favor of the adoption, based on the report submitted by the Social Worker of the Fifth Family Tribunal designated for this purpose, stating that the request be granted and a public deed recorded to this effect at the Office of Vital Statistics of this capital, as is proper. SECOND: The executing parties declare by this instrument that they confirm their will expressed in the aforementioned deed, by which the special judicial agent adopts, on behalf of the parties she represents, Mr. SAMUEL HUNTER and Ms. GRACE ELLIS, the minor JOSE CHAVEZ, as their own child, stating that because he has no assets, a relevant inventory was not drawn up. Each of the executing parties, declaring his or her own will in the capacity by which he or she appears, requests that this instrument be executed for the purpose of formalizing the adoption, and request that the Office of Vital Statistics of this capital proceed to annotate the birth certificate of the minor recorded by Number TWO THOUSAND THREE HUNDRED AND FORTY (2,340), page ONE HUNDRED AND FOUR (104), book ONE HUNDRED AND FIFTY (150) of births. THIRD: The executing parties state that they are aware of the rights and obligations they embark upon by this instrument, and the right of the adopted minor to use the surnames of his adoptive parents, Mr. SAMUEL HUNTER and Ms. GRACE ELLIS, to consider himself their own child, and to represent him before the law when necessary. FOURTH: Ms. CECILIA CHAVEZ, in accord with the relevant terms, accepts that her aforementioned minor child be adopted by the special judicial agent on behalf of Mr. SAMUEL HUNTER and Ms. GRACE ELLIS, and each of the executing parties, declaring his or her own will in the capacity by which he or she appears, expresses satisfaction with all and each of the clauses of this instrument. I, the Notary, attest to all that is stated, to having viewed the indicated deed for Notarial Adoption Proceedings, as well as the aforementioned opinion of the Attorney General's Office, and that by the authorization of the executing parties, have read all that is written for content, purpose, validity, and legal effect, and that in accord with the obligations of the Registration of the Testimony of this deed, they accept, authorize, and sign before the authorizing Notary.

Assignment 2: Notes

Note: This particular assignment presents ***many*** *examples of difficult terminology, but we only discuss the difficult ones in terms of dilemmas they may present. The rest of the terms that we do not point out specifically are not tricky per se, so we only mention their existence in order to underscore the impossibility of doing such an assignment without a dictionary.*

doy fe (line 3) - We highlighted this because countless times we run across documents where this standard phrase meaning *I attest* has been translated as *I give faith*. Of course, unless you work in a field where legal terms are used in both languages on a regular basis, you would not be able to translate this correctly unless you had a good legal dictionary, where under the word *dar*, you will find a huge list of "legalese" phrases involving all the legal definitions of the verb *to give* which exist.

por no serlo con anterioridad (line 5) - Again, unless the field of law is your area of expertise, this is an example of text that you may have to read several times in order to determine what exactly is being said. Then it is up to you to put that meaning into words which are in keeping with the style in which this document is written.

cédula de vecindad (line 5-6) - Obviously, every country has its own system for doing things. You have to do the research in order to determine what is meant when you run across a term such as this one which you have no reason to know offhand (nor could you always find it in a dictionary). This is where networking comes in. You will need to ask different people, including those at the embassy of the country where your document originates from in order to find out what this term means, and then you will have to coin a term that approximates it best and still makes sense to the reading audience.

municipalidad (line 7) and *municipio* (line 7) - The dictionary defines both of these in this context as *municipality*, but obviously, it would be silly to say *municipality of Salta, municipality of the Department of Belgrano*. So, we studied the text in further detail, did some research, and finally realized that in the second instance, *municipio* actually referred to the word in a jurisdictional sense, so it was not necessary to use it at all, and therefore we left out the second reference to *municipality* altogether.

Additional observations:

This being a technical, legal text, notice how the format of the translation follows closely that of the original. This includes capitalization of proper names and item numbers, spelling out numbers, and, in general, making the page closely resemble the original page.

Also, the original being one long paragraph with long sentences, great care was taken to make the English sentences sound clear and logical.

Assignment 3: Translation

Chemistry

Intercalation compound

Compound consisting of a crystalline lattice which acts as a donor of electrons and atoms which receive "strange" electrons which are intercalated or distributed throughout the planes of the lattice. An important group of intercalated compounds are those which are graphite, where bromium, for example, can act as an electron receiver. Graphite is particularly susceptible to this phenomenon because of its orderly stacks of crystal layers. Anhydric metallic nitrates such as copper and zinc nitrate also make up intercalated compounds with graphite. Another example is trilithium nitride, whose structure is made up of a series of layers of dilithium nitride, between which there is a layer of lithium atoms. This markedly increases conductivity such that the material is converted into a solid electrolyte effective in batteries. Other substances which have this property are titanium disulfide, and some metallic dioxides. The phenomenon does not alter the crystalline lattice and is reversible. The intercalated compounds are used as superconductors, synthetic lubricants, catalysts, and accumulators. They are also used in biochemical research; a compound based on acridine which can intercalate between pairs of bases of a DNA helix is used in cancer research.

Medicine

Allergies and Sensitization

Asthma, multiple chemical sensitivity, and skin diseases are on the rise in the work place. The classical definition of sensitization means a response initiated by the immune system. There is are difficulties in defining "atopy." Should a sensitizer be defined as that substance that can elicit an allergic response in the average person, or only in certain people?

Biological Monitoring

The production or inhibition of a particular metabolic compound or protein, or the cumulative damage to DNA, could provide a sensitive indicator of exposure to a toxic substance. Besides, genetic testing is becoming more sophisticated, cheaper, and more accessible to the average doctor. However, this raises ethical as well as scientific questions.

Cancer and Carcinogens

One problem that is emerging and causing uncertainty is how we identify and classify carcinogens. For example, recent changes to the ACOIH system of classification establish that an animal carcinogen for which epidemiology provides no evidence of its existence in human beings is an A3. This is contradicted, however, by the choice of substances to include in A3.

Chemicals and Toxicology

A number of interesting papers have been presented recently on toxicology, especially with regard to solvents and metals. There is also a trend toward greater interest in immunotoxicology (effects on the immune system) and neurotoxicology (effects on the nervous system).

Musculoskeletal Injuries (MSI)

Musculoskeletal injuries are now the single largest category of occupational disease and their frequency has increased about 20% annually in recent years. In analyzing MSI statistics, it is evident that solutions may be found by examining the changes that have taken place in work places during this time of rapid increase.

Genetics and Reproduction

There are three areas of concern: genetic and hormonal effects on reproduction, and effects on males.

Genotoxins - It is widely accepted that certain chemical products can cause genetic changes in humans that can in turn be passed on to children, causing disease. However, most people would be surprised to learn that despite years of research, there is no absolute scientific proof that similar consequences would result from exposure to any chemical product. Hormones - Recent reports have highlighted the possibility that there has been widespread contamination of the environment by compounds which disrupt the functioning of the endocrine system. Some scientists now believe that we have flooded the environment with compounds which contain estrogen.

Reproductive Hazards in Males - Most of the studies of adverse consequences on the reproductive system (birth defects, miscarriages, low birth weight) have focused on the pregnant woman's exposure to various hazards. What is less often studied is the effect the father's exposure and lifestyle may have on reproduction.

Mineral Fibers

Fiberglass and other synthetic mineral fibers are the main asbestos substitutes. The asbestos industry has made much of the fact that these substitutes themselves are not without risk. Certain lung diseases and cancers similar to those caused by asbestos have been observed in workers handling other fibers, but the evidence suggests that most are safer than asbestos.

Electromagnetic Fields and Other Radiation

All of us are continually surrounded by electromagnetic fields, both at work and at home. The results of several studies, although somewhat inconclusive and contradictory, tend to show a link between exposure to electromagnetic fields and certain types of cancer (especially tumors in the central nervous system and childhood leukemia).

Assignment 3: Notes

Note: Though this particular assignment presents ***many*** *examples of specific terminology, the text itself is straightforward, and presents no real difficulty in terms of sentence structure or language style. We therefore limit ourselves to explaining only those few points that to us represent dilemmas which may not be resolved just by using reference materials.*

capas de cristales ordenadamente apiladas (line 6) - At first glance, this can be translated any number of ways (even without a dictionary). *Layers of crystals in orderly stacks*, *orderly stacked layers of crystals*, etc. We chose *orderly stacks of crystal layers* not because it was that different than any of the other options, but because it was the least wordy and most efficient way of stating the phrase.

ACOIH (page 2, line 16) - This is an example of an acronym which remains the same in both texts because it pertains to a system of classification recognized as such in Spanish. As we have seen, names originally coined in one language only should be left as such, with the translation of the name in parentheses to let the reader know what it means, but in the rest of the text the corresponding acronym should be left in its original form. The same occurs with *lesiones musculoesqueléticas* (line 24, page 2) which has been left with its original *MSI* acronym.

la evidencia sugiere (page 3, line 19) - *It is evident* is a clear direct translation of this phrase. The literal *the evidence suggests* is not necessary, especially since the phrase follows the clause *Al analizar las estadísticas ...*

Additional observations:

A good chemical and/or medical dictionary should have most if not all of the technical terms in these two texts. If, after a thorough consultation with such sources you still have an unknown term, the next thing to do is call someone with a chemical or medical background.

Also, bear in mind that both chemistry and medicine have well defined vocabularies. Repeated work in any particular area of either will find you handling the same terms over and over again.

LOANS
and
TECHNICAL COOPERATION

TERMS AND CONDITIONS

The Bank's pool-based variable lending rate methodology, which was adopted in December 1989, uniformly adjusts the lending rate applicable to the outstanding balances of all loans approved since January 1, 1990. As a result, all borrowers equally share the costs and benefits of changes in the lending rate which is consistent with the cooperative nature of the institution. The lending rate obtained by applying the fixed-at-disbursement methodology is still applied to disbursements of loans approved prior to January 1, 1990. Both the variable and fixed-at-disbursement lending rates include the same spread over the relevant cost of borrowings, which in 1994 was 0.5 percent.

New borrowings are assigned to fund disbursements of loans approved under either of the two types of lending rates. The cost of borrowing for the fixed-at-disbursement rate is the weighted average cost of new borrowings carried out during the previous 12 months which are assigned to fund loans approved before January 1, 1990. The cost of borrowing for the variable lending rate is the weighted average for the previous six months of all outstanding borrowings assigned to the pool of borrowings that funds loans approved after January 1, 1990.

The lending rate for new disbursements of fixed-at-disbursement loans was 6.14 percent during the first six months of 1994 and 5.63 percent during the last six months of the year. The lending rate charged on outstanding balances of variable-rate loans was 6.95 percent during the first six months of 1994 and 6.81 percent during the last six months of the year. Loans extended from the ordinary capital during 1994 were made for terms ranging from 20 to 30 years.

During 1994, six loans were made from the ordinary capital's newly established U.S. dollar window, under which loans are denominated and disbursed only in U.S. dollars, with maturities of up to 10 years, and a choice of either a fixed rate or a LIBOR-based floating rate based on the Bank's cost of U.S. dollar funding of that type and the same lending spread as other ordinary capital loans. This program is aimed at stimulating private sector activity by providing financing on terms more compatible with the needs of private sector borrowers to whom the funds will be on-lent.

One loan was made in 1994 from the ordinary capital in the member's own currency. The financing carried an interest rate of 4 percent and a 20-year maturity.

The rates of interest charged on local currency loans from the Small Business Fund (SBF) varied from 1 to 4 percent depending on the stage of development of the country and the nature of the project, with grace periods ranging from five to ten years and maturities varying from 25 to 40 years. The rate of interest charged on convertible currency loans from the FSO for the less developed countries was 1 percent during the first ten years and 2 percent thereafter, with 10-year grace periods and 40-year maturities.

In addition to the interest rate on loans, the Bank charges a commitment fee of 0.75 percent on ordinary capital loans approved by January 1, 1989, 1.25 percent on ordinary capital loans approved prior to that date, and 0.50 percent on loans from the SBF, and a one-time inspection and supervision fee of 1 percent on all loans. As mentioned previously, 75 basis points of the latter fee were waived during the first semester of 1994.

Loans not extended directly to governments of the member countries, or to intergovernmental development institutions, carry guarantees of a government or of a government agency. The financing of small businesses, which need not be guaranteed, are extended in local currency loans with maturities of up to 40 years, with grace periods of up to 10 years, and are charged a commission of 1 percent per year. Technical cooperation is extended on a reimbursable, grant, or contingent recovery basis. Contingent recovery technical cooperation, which accounts for a small amount of the total, is subject to repayment only if as a result of the cooperation a loan is subsequently made by the Bank, or another external financial institution, for the execution of a project or program.

Assignment 4: Notes

Note: This text - in both Spanish and English—was originated by a major financial institution. It is quite difficult, and was chosen for that reason. Many times you will receive an assignment which may exceed your capability. You therefore have to learn to be realistic about your skills, and accept jobs accordingly.

The translated text may differ significantly from what you might have come up with, especially in regard to sentence structure. In this case, structure is especially important, because entire clauses become one word, and vice versa. These forewarnings are not meant to discourage you, but rather to show you the level of translation that could be expected from some of your clients. Obviously, in a case like this, the client may not hire a freelance translator, but would probably have this job done in-house. Your job would more likely be one of translating at a less technical level, of simply communicating the contents of the document to another person, but we included it in this set of exercises to demonstrate the higher end of the spectrum of the assignments that are out there, in terms of difficulty and required (and expected) skill level.

We will not highlight each and every point of difficulty, just those that are particularly representative of the problems presented by such a text.

tasa de interés (line 7) - Here, lack of knowledge of the subject matter could lead one to translate this as *interest rate* when, in fact, the text is discussing *lending rate*. This is crucial of course, and would only be apparent to someone who was familiar with the term *empréstitos* (line 14) in this context - that is, *borrowing* - which, when applied to related interest rates, refers to *lending* (or in this example, *lending rate methodology*).

se basa en una canasta (line 8) - The term *pool-based* makes perfect sense, once we see it in the translation! It shows up, however, at the beginning of the sentence as an adjective, whereas in the original it is part of a descriptive clause, although it still modifies the subject, *variable lending rate methodology*.

De esa manera, y en consonancia ... (lines 9-10) - The Spanish text, as is often the case with business texts of all sorts, including letters or reports, is more convoluted and uses more words than their translated English counterpart. In this case, this complicated phrase can be translated to its simple meaning *as a result*.

fijar la tasa en el momento del desembolso (line 12) - In English, particularly in the United States, it is more common to have phrases such as the translation of this example, *fixed-at-disbursement rate*, appear as ready-made hyphenated phrases, whereas in Spanish, they are written out in a way that actually explains what the phrase (or term) ***is*** rather than giving it a name of its own.

garantía (page 2, line 18) - The English translation is plural. Translating it in the singular would be understandable, but it would sound awkward because *guarantees* refers to those of the government ***or*** a government agency. Such circumstances are encountered often in both languages, so be aware of it. There are no set rules either from Spanish to English or vice versa. Words that fall into this category are numerous: law, statute, medicine, rights, etc. Again, using one over another is not necessarily incorrect, but rather is a matter of how it sounds.

Part Two

Translation: English into Spanish

Section 1. Assignments

5-7-/2 hours

Assignment 1: General

It is in the nature of humankind to explain things to themselves—how the earth was populated, where people originated, how long they have lived on the earth, how much longer they will be here. Most societies have origin myths or creation stories, migration legends, and histories. To better understand the Aztec peoples of the Fifth Sun, we must first look at what we estimate to be their actual origins.

The Bridge to the New World

The New World was new not only to the Europeans who discovered it in A.D. 1492. In geologic terms as well, the American continents were populated only recently. Most scientists agree that human beings, fully developed as a species called *Homo sapiens,* who had originated in Africa and had dispersed throughout the Old World, migrated to the New World at various times during the Pleistocene or Ice Age. This population movement was made possible by the formation of glaciers that lowered the sea level at certain periods between 30,000 and 12,000 B.C., allowing a land bridge to emerge between Siberia and Alaska. The basic assumption that men and women first entered the New World via the Bering Strait is widely accepted, although there is still disagreement about the specific migration routes they took and the date of their earliest appearance. Some groups of hunters may have migrated across the land bridge, called Beringia, following herds of big game—mammoths, mastodons, and bison. Others possibly followed routes across pack ice. Still other migrants may have rowed boats across the straits during the summer.

Beringia not only served as a migration route for people and herds from the Old World but also facilitated the departure of the horse and camel, which migrated to the Old World before dying out in the new. This event would have direct consequences for the New World inhabitants, since it would deprive them of the obvious benefit of large beasts of burden.

Archaeological evidence gathered from sites scattered throughout North, Central, and South America unfortunately has not provided clear, undisputed dates for the earliest occupations of New World peoples. Human populations had arrived at least by fourteen thousand years ago, and possibly much earlier. Some of the earlier radiocarbon dates, such as those gathered from sites in Mexico and South America, are still considered by many archaeologists to be problematic, however. Tlapacoya, a site near Mexico City, has radiocarbon dates of twenty-four thousand and twenty-one thousand years ago, and Peruvian sites have yielded dates as early as twenty thousand years ago. Whatever the exact time of their arrival, once in Alaska, small groups of hunters made their way south along the coast and through ice-free corridors to the northern Great Plains. From there they gradually penetrated other regions of the American continents. By 9000 B.C., a date accepted by most archaeologists, the New World was populated from Alaska to Tierra del Fuego, the southernmost tip of South America.

By the end of the Ice Age, around 10,000 B.C., gradual climatic warming produced environmental changes that profoundly affected the natural resources of the two American continents; equally dramatic as those which the planet will experience if the global warming recently forecast takes place. This early period of warmth may have contributed to the disappearance of mammoths and other large game in the Americas and caused a radical change in vegetation. As a consequence, early populations, called Paleo-Indians by archaeologists, were forced to elaborate new subsistence technologies, which involved a greater dependence on the hunting of small game, fishing, the gathering of shellfish, and the foraging for plants, wild grasses, nuts, seeds, fruits, and vegetables. Paleo-Indians eventually became more sedentary, making them more capable of exploiting particular environments more thoroughly.

The Birth and Growth of Societies

In the course of human dispersal over the span of two continents, different groups developed specific responses to the enormous diversity of environmental conditions. Some settled in extremely hostile natural settings where resources were scarce, necessitating seasonal migration in search of food. Others found conditions favorable enough to obtain abundant foodstuffs, thereby permitting more permanent settlements. With the passage of time, the more sedentary people became familiar with the life cycles and reproductive mechanisms of the plants they gathered and the animals they hunted. They eventually began to cultivate the plants to produce better, more resistant varieties. The intentional experimentation of early Native Americans, in conjunction with accidental mutations and other natural genetic change over many generations, gradually improved the quality and availability of preferred vegetable foods. The change from total reliance for survival on hunting and gathering to intensive foraging and occasional cultivation of specific plants marked the beginnings of agriculture in the New World. It would be many centuries, however, before permanent village life based solely on agriculture would make its appearance in the Americas.

word count: 807

4-5/1.5 hours

Assignment 2: Legal

CERTIFICATE OF INCORPORATION

OF

ABC, INCORPORATED

* * * * *

1. The name of the corporation is

ABC, INCORPORATED

2. The address of its registered office in the State of Delaware is 1234 Main Street, in the City of Essex, County of Sussex. The name of its registered agent at such address is the Registry Trust Company.

3. The nature of the business or purposes to be conducted or promoted is to engage in any lawful act or activity for which corporations may be organized under the General Corporation Law of Delaware.

4. The total number of shares of stock which the corporation shall have authority to issue is one hundred (100) and the par value of each of such shares is One Hundred Dollars ($100.00) amounting in the aggregate to ten Thousand Dollars ($10,000.00).

5A. The name and address of each incorporator is as follows:

5B. The name and address of each person, who is to serve as a director until the first annual meeting of the stockholders or until a successor is elected and qualified, is as follows:

6. The corporation is to have perpetual existence.

7. Elections of directors need not be by written ballot unless the by-laws of the corporation shall so provide. Meetings of stockholders may be held within or without the State of Delaware, as the by-laws may provide. The books of the corporation may be kept (subject to any provision contained in the statutes) outside the State of Delaware at such place or places as may be designated from time to time by the board of directors or in the by-laws of the corporation.

8. The corporation reserves the right to amend, alter, change or repeal any provision

contained in this Certificate of Incorporation, in the manner now or hereafter prescribed by statute, and all rights conferred upon stockholders herein are granted subject to this reservation.

9. A director of the corporation shall not be personally liable to the corporation or its stockholders for monetary damages for breach of fiduciary duty as a director except for liability (i) for any breach of the director's duty of loyalty to the corporation or its stockholders, (ii) for acts or omissions not in good faith or which involve intentional misconduct or a knowing violation of law, (iii) under Section 174 of the Delaware General Corporation Law, or (iv) for any transaction from which the director derived any improper personal benefit.

WE, THE UNDERSIGNED, being each of the incorporators hereinbefore named, for the purpose of forming a corporation pursuant to the General Corporation Law of the State of Delaware, do make this certificate, hereby declaring and certifying that this is our act and deed and the facts herein stated are true, and accordingly have hereunto set our hands this 10th day of September, 1990.

POWER OF ATTORNEY

For the granting of this Power of Attorney, Mr. Roy Sanders, appears in capacity of President of ABC, Inc., a company organized under the laws of Delaware, who being duly authorized by the Board of Directors of the Company, hereby grants a General Power of Attorney, which according to Law is broad and sufficient, in favor of Mr. Jack Smith, in order for him to act in name and representation of the Company to enter into negotiations with the Instituto Nacional de Información (hereinafter "INI") and others on behalf of ABC, Inc. and bind it with respect to the obligations which shall be performed in Venezuela in connection with a site management agreement to provide services to INI for maintenance, support, repair and operations of hardware and for the modification, support and operation of software for the modernization of the geographical information systems of INI.

In the event that Jack Smith becomes unavailable whether due to death, disability, illness or otherwise, then Joseph P. Roberts is hereby authorized to act on behalf of ABC, inc. instead of Jack Smith.

This Power of Attorney is valid for two (2) years from the date of signature.

WITNESS my hand and seal this sixteenth day of January, in the year of our Lord, one thousand nine hundred and ninety six.

word count: 677

6-8/2.5-3 hours

Assignmnet 3: Chemical/Medical

Chemistry

ARTICLE 9. Class 2 comprises compressed, refrigerated, liquefied, or pressure dissolved gases including substances which:

I. Have a vapor pressure of more than 300 kPa at 50 °C.
II. Are completely gaseous at 20 °C and at a standard pressure of 101.3 kPa.

For transportation conditions, substances in Class 2 are categorized according to their physical state as:

- Compressed gas, which, under pressure, is totally gaseous at 20 °C.
- Liquefied gas which is partially liquid at 20 °C.
- Liquefied refrigerated gas which is partially liquid because of its low temperature.
- Dissolved gas which is compressed and dissolved in a solvent.

Class 2 is divided as follows in keeping with the type of risk:

DIVISION DESCRIPTION OF SUBSTANCES

2.1 Flammable gases: Substances which, at 20 °C and a standard pressure of 101.3 kPa, burn when they are in a mixture of 13% or less by volume of air or have a range of flammability with air of at least 12% without reaching the lower flammability boundary.

2.2 Non-flammable, non-toxic gases: Gases that are transported at a pressure of no less than 280 kPa at 20 °C, or as refrigerated liquid gases and that:
  a) Are asphyxiating. Gases that dilute or replace the oxygen normally present in the atmosphere; or
  b) Are oxidant. Gases that, generally by releasing oxygen, can cause or contribute—more so than air — to the combustion of other material.
  c) Are not contained in the above.

2.3 Toxic gases: Gases which:
  a) Are known to be toxic or corrosive to human beings which is why they constitute a risk to health; or
  b) Are assumed to be toxic or corrosive to human beings because they have an LD of equal to or less than 5000 mol/m^3 (ppm).

Note: Gases meeting the above criteria due to their corrosiveness must be classified as toxic with a secondary risk of corrosion.

ARTICLE 10. Class 3 or flammable liquids. These are mixtures or liquids containing dissolved or suspended solid substances that release flammable vapors at a temperature of no more than 60.5 °C during closed-cup tests or no more than 65.6 °C in open-cup tests. Here are the substances included in this class:

- Liquids with an initial boiling point of equal to or less than 35 °C.
- Liquids with an inflammation point (in closed-cup) of less than 23 °C and an initial boiling point of more than 35 °C.
- Liquids with an inflammation point (in closed-cup) of greater than or equal to 23 °C, less than or equal to 60.5 °C, and an initial boiling point in excess of 35 °C.

ARTICLE 11. Class 4, flammable solids, are substances that present a risk of spontaneous combustion, as well as those that release flammable gases when in contact with water. They are subdivided as follows in keeping with the type of risk:

DIVISION	**DESCRIPTION OF SUBSTANCES**
4.1	Flammable solids: Solid substances that are not included among those classified as explosives but which, by virtue of the conditions prevailing during transportation, can easily burst into flame or can cause or accelerate fires due to friction.
4.2	Substances that present a risk of spontaneous combustion: Substances that can heat up spontaneously under normal transportation conditions or upon coming into contact with air and that can then burst into flame.
4.3	Substances that, in contact with water, release flammable gases: Substances that, due to reaction with water, can spontaneously become flammable or release flammable gases in hazardous quantities.

ARTICLE 12. Class 5, organic oxidants and peroxides, are substances that are defined and subdivided as follows in consideration of their risk:

DIVISION	**DESCRIPTION OF SUBSTANCES**
5.1	Oxidizing substances: Substances that, without necessarily being combustible, may generally cause or facilitate the combustion of others by releasing oxygen.
5.2	Organic peroxides: Organic substances that contain the bivalent structure O-O and that may be considered to be derived from hydrogen peroxide, where one of the hydrogen atoms, or both, were substituted by organic radicals. The peroxides are firmly unstable substances that can undergo self-accelerated exothermal decomposition. Besides, they may reveal one or several of the following properties: a) They may be susceptible to explosive decomposition; b) They may burn rapidly; c) They may be sensitive to shocks or friction; d) They may react dangerously on coming into contact with other substances; e) They may cause damage to vision.

ARTICLE 13. Class 6, acute (poisonous) toxics and infectious agents, these being substances that are defined and divided as follows in keeping with the risk:

DIVISION DESCRIPTION OF SUBSTANCES

6.1 Acute (poisonous) toxics: These are substances that can cause death, serious wounds, or that can be harmful to human health when ingested, inhaled, or when coming into contact with the skin. Compressed (poisonous) toxic gases can be included in the category of "Gases."

6.2 Infectious agents: These are agents that contain viable microorganisms, including bacteria, viruses, parasites, fungi, or a hybrid or mutant combination; these are substances that are known or that are believed to be capable of causing diseases in humans and animals.

ARTICLE 14. Class 7, radioactive substances, for purposes of transportation, includes all those materials whose specific radioactivity is in excess of 70 kBq/kg (2 nCi/g).

Emphysema is an abnormal condition of the lungs in which there is overinflation of the alveoli, leading to a breakdown in their walls, a decrease in respiratory function, and, in severe cases, increasing breathlessness. The disease appears to be associated with chronic bronchitis, smoking, and advancing age; one form that occurs early in life is related to a hereditary lack of an enzyme. Early symptoms of emphysema include dyspnea, cough, and rapid heart rate. Advanced cases are marked by signs of oxygen lack (restlessness, weakness, confusion) and frequently by complications of pulmonary edema and congestive heart failure. Breathing exercise, bronchodilators, and the prevention of respiratory infections may be helpful in treating the disease; however, severe cases may require oxygen.

Encephalitis is an inflammation of the brain, usually due to viral infection but which sometimes occurs as a complication resulting from another infection (e.g., influenza, measles) or resulting from poisoning. Symptoms include headache, drowsiness, neck pain, nausea, and fever, followed sometimes by neurologic disturbances such as seizures, paralysis, and personality changes. The outcome depends on the cause, extent of the brain inflammation, and general condition of the patient.

word count: 1014

3-4/1-2 hours

Assignment 4: Business/Finance

Value Portfolio

Creates Wealth Over the Long Term

Creating wealth over the long term with moderate risk is the primary objective of our Value Portfolio. And with an 18.6% compounded annual return (net of all fees) since its inception in 1974, it has a proven track record for meeting client goals and is Compu-Val's greatest success story.

In selecting stocks for our Value Portfolio, we follow a conservative investment principle, purchasing stocks that are inexpensive relative to the value of assets that the company controls. To do that, we calculate a **Value/Cost** ratio for each company, determining the market value of debt plus equity—the Value component—and comparing it to the company's inflation adjusted net assets—the Cost component.

In addition, we purchase only stocks of profitable companies with improving corporate performance as measured by cash flow return on investment (CFROI). Focusing on CFROI eliminates accounting distortions and give a true "inflation adjusted" measure of corporate performance. The accurate comparisons we are able to make between companies lead to exceptional stock performance.

When evaluating a stock for purchase, we look for firms that combine low debt and financial strength. We favor companies with high quality products or services, particularly in niche markets, that offer attractive long-term opportunities. In addition, we buy companies that grow at a sustainable rate and focus on core businesses.

Portfolio holdings are monitored daily and reviewed each week by Compu-Val's investment professionals. When a stock's market price reaches our goal, we sell. When a stock has a loss of 15% or more versus average cost, it is reviewed by our Investment Committee.

Fixed Income Portfolio

Minimizes Risk for Conservative Investors

Our Fixed Income Portfolio was established in 1984 to provide an attractive option for more conservative, risk-averse investors. It is designed to obtain the maximum return consistent with limited principal fluctuations and minimum credit risk. With a compounded annual return rate of 9.6% (net of fees) since its inception, our Fixed Income Portfolio has provided investors with consistent and positive results each year. Investments are limited to Government bonds and corporate debt instruments rated "A" or better by Standard and Poor's/Moody bond rating services. Maturities are staggered, with the average portfolio maturing in 6 years or less. Investments are chosen based on our analysis of premiums for risk taken versus risk-free Government securities.

word count: 393

Assignment 1: Translation

Es la naturaleza de los seres humanos explicarse cosas a sí mismos: cómo se pobló el globo, de dónde originaban las personas, cuánto tiempo han vivido sobre la Tierra, cuánto más durarán. La mayoría de las sociedades tienen mitos sobre su origen o narrativas acerca de la creación, leyendas de migraciones, e historias. Para comprender mejor a los pueblos aztecas del Quinto Sol, debemos primero mirar lo que estimamos sean sus orígenes reales.

El puente hacia el Nuevo Mundo

El Nuevo Mundo no fue nuevo sólo para los europeos que lo descubrieron en 1492. Además, en términos geológicos, los continentes americanos fueron poblados sólo recientemente. La mayoría de los científicos están de acuerdo en que los seres humanos, plenamente desarrollados como la especie llamada *Homo sapiens*, originada en Africa y dispersada por el Viejo Mundo, migró al Nuevo Mundo en varias épocas durante el pleistoceno o Edad Glacial. Este movimiento de población fue posible por la formación de glaciares que hicieron descender el nivel del mar en ciertos períodos entre 30.000 y 12.000 años A. de C., permitiendo que emergiera un puente de tierra entre Siberia y Alaska. El supuesto primordial de que hombres y mujeres penetraron primero al Nuevo Mundo por el estrecho de Bering es aceptado ampliamente, aunque todavía existe desacuerdo acerca de las rutas específicas de las migraciones y de la fecha de su primera aparición. Algunos grupos de cazadores pueden haber emigrado cruzando el puente de tierra, llamado Beringia, siguiendo manadas de caza mayor — mamuts, mastodontes, y bisontes. Otros posiblemente siguieron rutas a través de la masa de hielo. Aun otros migratorios pueden haber remado sus botes a través de los estrechos en la época del verano.

Beringia no sólo sirvió como una ruta migratoria para la gente y los rebaños del Viejo Mundo, sino facilitó también la partida del caballo y del camello que migraron al Viejo Mundo antes de extinguirse en el Nuevo. Este hecho tendría consecuencias directas para los habitantes del Nuevo Mundo, puesto que les privaría del obvio beneficio de tener grandes bestias de carga.

La evidencia arqueológica recogida de lugares diseminados por Norte, Centro y Suramérica lamentablemente no ha proporcionado fechas claras e irrefutables acerca de las primeras ocupaciones de los pueblos del Nuevo Mundo. Las poblaciones humanas habían llegado hacía por lo menos catorce mil años y posiblemente mucho antes. Sin embargo, algunas de las fechas anteriores registradas por radiocarbono, tales como las recogidas de lugares de México y Suramérica, todavía se consideran problemáticas por muchos arqueólogos. Tlapacoya, un lugar cerca de la ciudad de México, registra fechas de radiocarbono de hace veinticuatro mil y veintiún mil años atrás, y lugares en el Perú han producido fechas hasta de hace veinte mil años atrás. Cualquiera que haya sido el momento exacto de su llegada, una vez en Alaska, pequeños grupos de cazadores abrieron camino hacia el sur por la costa y a través de los corredores libres de hielo, hasta las grandes praderas del norte. Desde allí, gradualmente penetraron otras regiones de los continentes americanos. Para el año 9.000 A. de C., una fecha aceptada por la mayoría de los arqueólogos, el Nuevo Mundo estaba poblado desde Alaska hasta Tierra del Fuego, el extremo más austral de Suramérica.

A fines de la Edad Glacial, alrededor del año 10.000 A. de C., el calentamiento gradual del clima produjo cambios ambientales que afectaron profundamente los recursos naturales de los dos continentes americanos, igualmente dramáticos que los que va a experimentar el planeta si llega a producirse el calentamiento global recientemente pronosticado. Este temprano período de calor pudo haber contribuido a la extinción de los mamuts y otras especies de caza mayor en las Américas y haber ocasionado cambios radicales en la vegetación. A consecuencia de ello, poblaciones antiguas, llamadas paleoindias por los arqueólogos, se vieron forzadas a elaborar nuevas técnicas de subsistencia que implicaron una mayor dependencia de la caza menor, pesca, recolección de mariscos, y búsqueda de plantas, pastos silvestres, nueces, semillas, frutas, y vegetales. A la larga, los paleoindios se volvieron más sedentarios, lo cual mejoró su capacidad para explotar ciertos ámbitos de forma más completa.

El nacimiento y crecimiento de las sociedades

En el curso de la dispersión humana sobre el espacio de dos continentes, grupos diferentes desarrollaron respuestas específicas a la enorme diversidad de las condiciones ambientales. Algunos se asentaron en ambientes naturales extremadamente hostiles, donde los recursos eran escasos, exigiendo migraciones estacionales en búsqueda de alimentos. Otros hallaron condiciones suficientemente favorables para obtener abundantes víveres, así permitiendo asentamientos más permanentes. Con el paso del tiempo, las poblaciones más sedentarias se familiarizaron con los ciclos de vida y mecanismos reproductivos de las plantas que recogían y los animales que cazaban. Finalmente comenzaron a cultivar las plantas para producir variedades mejores y más resistentes. La experimentación intencional de los primeros aborígenes americanos, junto con mutaciones accidentales y otros cambios genéticos naturales a través de muchas generaciones, gradualmente mejoró la calidad y disponibilidad de los alimentos vegetales preferidos. El cambio de una total dependencia de la caza y la recolección para subsistir, al forrajeo intensivo y cultivo ocasional de plantas específicas, marcó los comienzos de la agricultura en el Nuevo Mundo. Sin embargo, transcurrirían muchos siglos antes de que hiciera su aparición en las Américas una vida aldeana permanente, basada sólo en la agricultura.

Assignment 1: Notes

earth (line 1) - There are several translations for this, but we chose *globo* because in the original, the term covered earth in the sense of the world. *Tierra* would also have been a good choice, but we needed another synonym for earth in the same sentence, so we used *tierra* for that one.

Aztec (line 4) - In Spanish, as you know, the rules for capitalizing certain terms are very different than in English. Months, days of the week, nationality, origin, religion, and the like are ***never*** capitalized. Proper names, such as names of countries or of people are. Although this may seem obvious to some, there are people who after living in the United States for a while, begin to adopt the rules that govern proper English spelling and format, even when writing in Spanish.

actual (line 5) - This is a good example of a word that if treated as a cognate will inevitably lead to error. In English, *actual* means true or real. In Spanish it means current or present. So in this case, we have to translate actual as *real.*

A.D. 1492 (line 7) - In Spanish, a date standing alone is assumed to be *A.D.*, even in the context of a historical description such as this one. If it is B.C., it is translated as **A.C.** There is no real translation for *A.D.*, and therefore the differentiation is only made when referring to B.C.

earliest appearance (line 14-15) - We have highlighted this to illustrate the difference between early in terms of time of the day and early as it refers to chronology. Early in the day is *temprano. Earliest appearance* is actually a fancy way of saying first appearance, so it is translated as *primera aparición. Aparición más temprana* would simply be an example of poor Spanish. *Earlier* as it appears later on (line 26) is translated as *anterior* since it refers to a previous date, as opposed to a first date. Again, *early*, as it appears later on in the text (line 4, page 2), is translated as *temprano. Early*, as in *early populations* (line 6, page 2) or *early Native Americans* (line 19, page 2) has been translated as *antigua* and *primeros*, because of the complete difference in context.

pack ice (line 17) - *Pack ice* is not your average everyday term, so some research is needed to find out what it is in order to be able to translate it correctly as *masa de hielo.*

migrated (line 20) - The first instinct would be to translate this as a conjugation of the verb *migrar* when actually the correct translation is *emigrar. Migrar* sounds right, looks right, and is often used, but it is still wrong. The same occurs with *migrants* (line 17). *Migrante* looks and sounds right, but, surprisingly, the proper term is *migratorio*, which seems like it should mean *migratory*. This is a good reminder to not take any translation for granted and to ***always*** use a dictionary.

clear, undisputed (line 24) - Using two adjectives together, separated only by a comma, does not work in most types of Spanish text, so instead they have to be translated with an "and" in between, which still retains the author's meaning and is in keeping with the rules for good Spanish. The same occurs with the translation of *better, more resistant* (page 2, line 23) which resulted in *mejores y más resistentes*.

however (line 28) - This is an example of changing the order within the sentence. In English, the word *however* can be placed in many parts of the sentence and remain correct, as well as convey the same sense in the phrase. The position of the term *sin embargo*, on the other hand, is less flexible in the same instance, especially because there is another clause in the middle describing where the radiocarbon data were gathered from. In order to keep *sin embargo* from modifying the wrong part of the sentence or from making the sentence sound awkward, it really can only go in the beginning. The same occurs with the sentence *It would be many centuries, however* (lines 28-29, page 2).

has (line 28) - *Has* could easily have been translated as *tiene*, but in this case, the use of the word has implies a demonstration or proof of something, so *registra* is more accurate and conveys the idea that radiocarbon dating is something which is recorded and which clearly proves something. *Tiene*, would not have had the same impact, but would be a much more vague, albeit correct, way of writing this in Spanish.

made their way (page 2, line 1) - This phrase could have been translated in numerous ways. We chose *abrieron camino* as the best way of trying to keep to the sense that these were explorers, pioneers, breaking into uncharted, unknown territory, "clearing a path" for themselves.

southernmost (line 4) - *Punta más al sur* is technically correct, but it is not a geographic term, so to speak. In Spanish the cardinal point south is *sur*, but in this type of description, it is more correct to refer to this region as *austral*.

; (line 6) - The author used a semicolon here which was not actually necessary in English, and much less so in Spanish.

global warming (line 7) - This is another case where research pays, especially in terms of keeping up with publications such as newspapers or magazines, where you will be able to find out what term is used for *global warming (calentamiento global)* as far as common usage is concerned. There are many phrases such as this which have been commonplace in the United States, for example, for some time, and have recently made their way into Spanish, and therefore may not necessarily be in any dictionary, especially if they have been coined by the media.

large game (line 9) - The translation *especies de caza mayor* could just as easily have read simply as *caza mayor*. This is an illustration of using judgment and instinct for what sounds better, yet not changing the author's message in any way.

eventually (page 2, lines 13 and 22). It was translated differently in each case because of the context. In the first instance it was better to say *a la larga* because this conveys the idea that there it was the result of a progression of events. In the second, *finalmente* implies that this is the end of the sequence of events. The position this word appears in these sentences is again up to the translator in terms of what sounds best. This is not an issue of absolutes, since what is being weighed here is not the use of the term, but the way the sentences flow best. As you no doubt realize, the success of a translation is unquestionably based on accuracy, but without good instincts for sound writing in both languages, there will be many occasions when you will get "stuck" not for want of a correct term, but rather for not knowing how to create the best structure for the sentence as a whole.

to intensive foraging ... specific plants (line 26-7) - This clause, when translated into Spanish, had to be separated from the rest of the sentence by commas so it would be clear.

ESCRITURA DE CONSTITUCION

DE

ABC, INCORPORATED

* * * * *

1. El nombre de la sociedad es

ABC, INCORPORATED

2. La dirección de su domicilio social en el Estado de Delaware es 1234 Main Street, en la ciudad de Essex, Condado de Sussex. El nombre de su agente acreditado en esa dirección es Registry Trust Company.

3. La índole del negocio o de los propósitos que se llevarán a cabo o se promoverán es la realización de actos o actividades dentro de la ley para las cuales las sociedades pueden organizarse bajo la Ley General de Constitución de Sociedades de Delaware (General Corporation Law of Delaware).

4. El número total de acciones que la sociedad tendrá derecho de emitir es de cien (100), y el valor nominal de cada una de esas acciones es de Cien Dólares ($100), sumando a un total de Diez Mil Dólares ($10.000).

5A. Nombre y dirección de cada gestor:

5B. Nombre y dirección de cada persona que va a servir de director hasta que tenga lugar la primera asamblea anual de accionistas, o hasta que un sucesor sea elegido y habilitado:

6. La sociedad tendrá vida perpetua.

7. Las elecciones de los directores no tienen que ser por voto escrito, a menos que los estatutos de la sociedad lo requieran. Las reuniones de accionistas pueden constituirse adentro o afuera del estado de Delaware, según lo dicten los estatutos. Los libros de la sociedad pueden guardarse (sujeto a cualquier disposición que figure en los estatutos) afuera del estado de Delaware en el sitio o los sitios que sean designados periódicamente por la junta directiva o de acuerdo a los estatutos de la sociedad.

8. La sociedad se reserva el derecho de enmendar, alterar, cambiar o revocar toda disposición contenida en esta Escritura de Constitución de la manera prescrita ahora o en adelante de acuerdo con los estatutos, y todo derecho conferido aquí a los accionistas se otorga bajo esa reserva.

9. El director de la sociedad no será personalmente responsable ante la sociedad o los accionistas por daños monetarios causados por incumplimiento de su responsabilidad fiduciaria como director, pero retendrá la responsabilidad (i) por cualquier violación de su obligación de lealtad ante la sociedad y los accionistas, (ii) por actos u omisiones que no sean en buena fe o que entrañen mala conducta intencional o una violación adrede de la ley, (iii) en virtud de la sección 174 de la Ley General de Constitución de Sociedades de Delaware, o (iv) por cualquier operación en que el director haya derivado algún beneficio personal impropio.

NOSOTROS, LOS ABAJO FIRMANTES, siendo los gestores arriba nombrados, a fines de formar una sociedad según la Ley General de Constitución de Sociedades del Estado de Delaware, hacemos la presente escritura por la cual declaramos y certificamos que es nuestro acto e instrumento y que los hechos aquí expuestos son verdaderos, por lo que la hemos firmado en este día, el 10 de septiembre de 1990.

PODER NOTARIAL

A fin de otorgar este Poder Notarial, el señor Roy L. Sanders comparece en su capacidad de Presidente de ABC, Inc., una compañía constituida según las leyes de Delaware, y estando debidamente autorizado por la Junta Directiva de la Compañía, por la presente otorga un Poder Notarial, que por ley es amplio y suficiente, al señor Jack Smith, para que actúe en nombre de la Compañía y como su representante, a fin de iniciar negociaciones con el Instituto Nacional de Información (que llamaremos "INI") y otros, en nombre de ABC, Inc., y vincularlo respecto a las obligaciones que se efectuarán en Venezuela en relación con un acuerdo de administración de un sitio para proveer servicios a INI para el mantenimiento, apoyo, reparación y utilización de computadoras y para la modificación, apoyo y utilización de programas de computación para la modernización de los sistemas de información geográfica de INI.

En caso de que Jack Smith fuera inhabilitado por muerte, invalidez, enfermedad, u otro motivo, por la presente se autoriza a Joseph P. Roberts actuar en nombre de ABC, Inc. en lugar de Jack Smith.

Este Poder Notarial es válido por dos (2) años, a partir de la fecha de su firma.

Mi firma y sello dan fe en este día dieciséis de enero de mil novecientos noventa y seis.

Assignment 2: Notes

Certificate of Incorporation (line 1) - At first glance, the natural impulse would be to translate this as *Certificado de Incorporación*, but translation of legal documents can be very tricky, as we will explain. Although any person in the business world from any Spanish-speaking country will know what the term *incorporar* or *corporación* (line 5) means thanks to previous exposure in dealing with businesses based in the United States, these terms are very different in most of these countries. The best rule of thumb is to write for the audience, not for the person who hired you. In other words, you must keep in mind that the whole purpose of translation is not to create, but to communicate, and to do this in the most accurate way possible. You are not an author, but a messenger.

In this case, therefore, you need to use the terms that your audience normally uses, even if they could understand the not-so-correct ones. Thus, in this example, we need to refer to *incorporation* as its true meaning, which is to make up the *corporation*, in this case, *sociedad* (or *sociedad anónima*). The same goes for all other related terminology such as *incorporator* (line 16) which is most correctly translated as *gestor*. Of course, as you will get tired of being told, without proper reference materials, such as legal and business/commercial dictionaries, this task of finding the right words will be impossible.

ABC, INCORPORATED (line 6) - Here there is no need to translate the name of the company, because it would not mean anything in Spanish. It will not be listed anywhere as *COMPAÑIA ABC* or *SOCIEDAD ABC* and the name has no significance in and of itself. The same is true in the case of *Registry Trust Company* (line 9) and *General Corporation Law of Delaware* (line 11 - 12). We still put the translation *Ley General de Constitución de Sociedades* in parentheses so the reader will know what it means, but we are not rewriting the names of the statutes of the state of Delaware.

registered office (line 7) - This is yet another excellent example of how such a translation (i.e., legal) *requires* a dictionary. Without this, you might be tempted to translate this as "oficina registrada" when, in fact, what we are actually discussing here is the headquarters of the company, hence *domicilio social*. The same occurs in the case of *breach of fiduciary duty* (page 2, line 2).

registered (line 8) - Again, although *registrado* naturally springs to mind, a dictionary will give you *acreditado* or another such synonym. Accuracy, not assumptions, is the key.

nature (line 10) - Any number of words would have been appropriate here, such as *naturaleza*, *clase*, *tipo*, and the like. *Indole*, however, was deemed the best choice because of the legal context. This is a matter of judgment and interpretation of the author's style, nothing more.

lawful (line 11) - The spirit of the definition or translation of *lawful* is "within the law." Therefore, rather than use a literal translation such as *legal*, we use *dentro de la ley*. Such use of words is also well within the style of the legal language used in the text.

($100.00) and ($10,000.00) (lines 14 and 15) - In Spanish, when an amount of money is referred to in American currency, it is expressed only in dollars if no cent amount is mentioned; so, in this case, it was not necessary to include the .00 in the translation of these amounts. If there had been any amount in cents, then we would have translated it as *$100 con 50 centavos*, for example. To write it out as we would in English would be very confusing to some, since the use of the comma and decimal point in Spanish is the opposite of that in English. This would be confusing to all, regardless of their mother tongue or country of residence.

is as follows (line 16) - There is no need to translate *is as follows*; this is conveyed by the use of the colon. To actually state it would make the sentence sound awkward, i.e., *es como sigue*. The same occurs with *The name and address of each person ...* (line 17-18).

provide (line 21) - The word *provide* has many possible translations, since it has many synonyms. As we have explained before regarding avoiding being too literal, in translating this term we should look to the essence of what is being conveyed - that is, the laws *provide* really means *los estatutos dictan*. A law does not literally *provide* anything, but rather governs, rules, or states. The same thing occurs with the use of the word *in* (line 23) where the real meaning is that this idea is *contained* within the law, according to the law.

amend, alter, change or repeal (line 25) - These all being synonyms, you have to be especially careful in picking the right ones. Sound judgment combined with a good dictionary is what this exercise requires.

all rights (page 2, line 2) - The Spanish translation is singular. Be careful not to forget your "good" Spanish in situations like these. To translate it in plural would make it sound cumbersome. Such circumstances are encountered often in both languages, so be aware of it. There are no set rules either from Spanish to English or vice versa. Words that this happens with are numerous: law, statute, medicine, rights, etc. Again, using one over another is not necessarily incorrect, but rather is a matter of how it sounds.

power of attorney (page 2, line 15) - Again, in many Spanish-speaking countries, legal proceedings are carried out by *notarios* which may translate literally as *notaries*, but which has a very different meaning in those countries than it does in the United States where notaries only attest that a document is valid. *Notarios* or, in some countries, *escribanos*, actually carry out many legal transactions and proceedings (some ***are*** lawyers) such as wills and trusts, adoptions, certificates of incorporation, deeds, contracts, and other agreements. Hence, *Power of Attorney* is *Poder Notarial*. It is very important in translation to be aware of and knowledgeable about such cultural differences in what may seem to be standard procedures and activities.

in the year of our Lord (line 29) - In Spanish, a date standing alone is assumed to be *A.D.*, even in the context of a historical description such as this one. If it is B.C., it is translated as *A.C.* Since there is no real translation for *A.D.*, the differentiation is only made when referring to B.C.

Note: At the end of the original document, it mentions that there is a seal. Many of these sorts of documents have seals, stamps, franks, and signatures which have to be translated as well. You must refer to the position of the item, followed by the translation, and enclose the whole item in brackets. For example: [seal in bottom left hand corner: signature, September 17, 1967] on the page where the seal appears.

Chemistry

ARTICULO 9 - La Clase 2 comprende gases comprimidos, refrigerados, licuados o disueltos a presión, incluyendo sustancias que:

I. Tienen una presión de vapor mayor superior a 300 kPa a 50 °C.
II. Son completamente gaseosas a 20 °C a una presión normal de 101,3 kPa.

En cuanto a las condiciones de transporte, las sustancias de Clase 2 se clasifican de acuerdo con su estado físico tal como:

- Gas comprimido que bajo presión es totalmente gaseoso a 20 °C.
- Gas licuado que es parcialmente líquido a 20 °C.
- Gas licuado refrigerado que es parcialmente líquido debido a su baja temperatura.
- Gas disuelto que está comprimido y disuelto en un solvente.

La Clase 2 se divide según el tipo de riesgo en lo siguiente:

DIVISION DESCRIPCION DE LAS SUSTANCIAS

2.1 Gases inflamables: Sustancias que a 20 °C y una presión normal de 101,3 kPa, arden cuando se encuentran en una mezcla de 13% o menos de aire por volumen o en una gama de inflamabilidad con aire de al menos 12% sin alcanzar el límite inferior de inflamabilidad.

2.2 Gases no inflamables, no tóxicos: Gases que son transportados a una presión no menor de 280 kPa a 20 °C, o como gases líquidos refrigerados y que:
  a) Son asfixiantes. Gases que diluyen o reemplazan el oxígeno normalmente presente en la atmósfera; o
  b) Son oxidantes. Gases que generalmente mediante la liberación de oxígeno, pueden causar la combustión de otro material o contribuir a ella más que el aire.
  c) No están contenidos en los anteriores.

2.3 Gases tóxicos. Gases que:
  a) Se conocen como tóxicos o corrosivos para los seres humanos por lo cual constituyen un riesgo para la salud; o
  b) Se supone que son tóxicos o corrosivos para los seres humanos porque tienen un CL igual a o menor que 5.000 Mol/M³ (ppm).

Nota: Los gases que cumplen con los criterios anteriores debido a su corrosividad, deben clasificarse como tóxicos con un riesgo secundario de corrosión.

ARTICULO 10 - La Clase 3 o líquidos inflamables. Estos son mezclas o líquidos que contienen sustancias sólidas en solución o suspensión que liberan vapores inflamables a una temperatura
no superior a 60,5 °C en los ensayos en copa cerrada o no superiores a 65,6 °C en ensayos de copa abierta. Aquí siguen las sustancias incluidas en esta clase:

- Líquidos con un punto de ebullición inicial igual a o menor que 35 °C.
- Líquidos con un punto de inflamación (en copa cerrada) menor que 23 °C, y un punto inicial de ebullición mayor que 35 °C.
- Líquidos con un punto de inflamación (en copa cerrada) mayor que o igual a 23 °C, menor que o igual a 60,5 °C, y un punto inicial de ebullición mayor que 35 °C.

ARTICULO 11 - La Clase 4, sólidos inflamables, son sustancias que presentan un riesgo de combustión espontánea, así como aquellos que en contacto con el agua liberan gases inflamables. Se subdivide según el tipo de riesgo en lo siguiente:

DIVISION DESCRIPCION DE LAS SUSTANCIAS

4.1 Sólidos inflamables: Sustancias sólidas que no están incluidas entre las clasificadas como explosivas pero que en virtud de las condiciones que se dan durante el transporte, pueden inflamarse con facilidad o pueden provocar o acelerar incendios a causa de fricción.

4.2 Sustancias que presentan un riesgo de combustión espontánea: Sustancias que pueden calentarse espontáneamente bajo condiciones normales de transporte o al entrar en contacto con el aire y que entonces pueden inflamarse.

4.3 Sustancias que en contacto con el agua desprenden gases inflamables: Sustancias que por reacción con el agua pueden volverse espontáneamente inflamables o desprender gases inflamables en cantidades peligrosas.

ARTICULO 12 - La Clase 5, oxidantes y peróxidos orgánicos, son sustancias que se definen y subdividen según su riesgo en lo siguiente:

DIVISION DESCRIPCION DE LAS SUSTANCIAS

5.1 Sustancias oxidantes: Sustancias que sin ser necesariamente combustibles, pueden, generalmente mediante la liberación de oxígeno, provocar o facilitar la combustión de otras.

5.2 Peróxidos orgánicos: Sustancias orgánicas que contienen la estructura bivalente O-O y que pueden considerarse derivados del peróxido de hidrógeno, en que uno de los átomos de hidrógeno, o ambos, han sido sustituidos por radicales orgánicos. Los peróxidos son sustancias firmemente inestables que pueden sufrir una descomposición exotérmica autoacelerada. Además, pueden tener una o varias de las siguientes propiedades:
a) Ser susceptibles a una descomposición explosiva;

b) Arder rápidamente;
c) Ser sensibles a los impactos o a la fricción;
d) Reaccionar peligrosamente al entrar en contacto con otras sustancias;
e) Causar daño a la vista.

ARTICULO 13 - La Clase 6, tóxicos agudos (venenos) y agentes infecciosos; estas sustancias se definen y dividen según su riesgo en lo siguiente:

DIVISION	DESCRIPCION DE LAS SUSTANCIAS
6.1	Tóxicos agudos (venenosos): Estas son sustancias que pueden causar la muerte, lesiones graves, o que pueden ser dañinas para la salud humana si se ingieren, se inhalan, o si entran en contacto con la piel. Los gases tóxicos (venenosos) comprimidos se pueden incluir en la categoría de "Gases".
6.2	Agentes infecciosos: Estos son agentes que contienen microorganismos viables incluyendo bacterias, virus, parásitos, hongos, o una combinación híbrida o mutante; estas sustancias se conocen como o se cree que pueden ser capaces de provocar enfermedades en los seres humanos y los animales.

ARTICULO 14 - La Clase 7, sustancias radiactivas, para los efectos de transporte, incluye todos los materiales cuya radioactividad específica es superior a 70 kBq/kg (2 nCi/g).

El enfisema es una condición anormal en la cual hay sobreinflación de los alvéolos, lo cual lleva a la destrucción de sus paredes, reducción de la función respiratoria y, en casos muy severos, un aumento en la falta de aire. La enfermedad parece estar relacionada con la bronquitis crónica, con el fumar, y con el envejecimiento; un tipo de enfisema que ocurre temprano en la vida está relacionado con la deficiencia hereditaria de un enzima. Los síntomas tempranos del enfisema incluyen la disnea, la tos, y la aceleración del ritmo cardíaco. Los casos avanzados se destacan por señales de falta de oxígeno (inquietud, debilidad, confusión) y frecuentemente por complicaciones debidas al edema pulmonar y a la insuficiencia cardíaca congestiva. Los ejercicios para la respiración, los broncodilatadores, y la prevención de infecciones respiratorias pueden ser beneficiosos en el tratamiento de la enfermedad; sin embargo, los casos severos pueden requerir oxígeno.

La encefalitis es una inflamación del cerebro generalmente ocasionada por una infección viral, aunque a veces ocurre debido a una complicación de otra infección (por ejemplo, la gripe o el sarampión) o al envenenamiento. Los síntomas incluyen el dolor de cabeza, la soñolencia, el dolor del cuello, las náuseas y la fiebre, a veces seguidos por perturbaciones neurológicas tales como ataques, parálisis, y cambios de personalidad. El pronóstico depende de la causa, de la gravedad de la inflamación cerebral, y de la condición general del paciente

comprises (line 4) - There are several translations for this, but we chose *comprende* because in the original, the term is used in a technical sense - it groups a number of elements into a specific category. This is a good opportunity to point out that in certain technical documents, the same word will be used over and over, and though you may not realize it, it is very easy to forget (especially when working on long texts) which word you used. Rather than risk making the translation inconsistent, or waste time going back and forth through the document to see which word you chose, it is best to write down the terms that will be repeated on an index card, for example, and use this as a quick reference guide.

compressed gas (line 10) - You cannot accept an assignment such as this one without the proper reference tools. Unless you work in the science field in a bilingual environment, it is not possible to know all the correct scientific terminology, since it is not exactly dinner table conversation, nor is it likely to be published often enough in a newspaper or magazine for you to be exposed to it on a daily basis. The same occurs with the terms *liquefied gas* (line 11), *liquefied refrigerated gas* (line 12), *dissolved gas* (line 13), and *volume of air* (line 17).

LD (line 29) - Apparently, *LD* (*lethal dose*), means something very basic to the author in its abbreviated format since the full term was not provided along with it. Some such abbreviations appear in the dictionary, in which case you are fine, but when they do not, then some detective work needs to be done. In this case, a technical and/or chemical dictionary gives two possibilities, one being length to distance ratio, an engineering term, and the other being *lethal dose*. Using your judgment and based on the context, you would use the translation for *lethal dose* (*concentración letal*) and abbreviate it to *CL*.

inflammation point (page 2, line 6) - This represents something which happens with some frequency, namely that a term appears in the original text which is not correct. There is no *inflammation point* in English, but the context would appear to indicate that the author meant *point of flammability* which is *punto de inflamabilidad*. On another note altogether, bear in mind that in Spanish, and often in English, *flammable* and its derivatives have been corrupted for public use to *inflammability* and ***its*** derivatives, in order to avoid confusion when these words are used in instructions or warnings.

oxidizing substances (line 27) and *organic peroxides* (line 28) - These are further illustrations of technical terms that should be looked up, not guessed at.

one form (page 4, line 5) - Although *una forma* would be understood, it is not as formally correct as *un tipo*.

of (line 8) - The word *of* can be translated any number of ways, depending, again, on context. Here, the author is saying *complications of*, meaning that something is causing, or is caused by, a certain factor, so *debido a*, *owing to*, fits best in the translation.

; however (line 10) - This is an example of changing the structure and punctuation within the sentence. In English, as we have seen, the word *however* can be placed in many parts of the sentence and remain correct, as well as convey the same sense in the phrase. In this case, if the phrase *sin embargo* were in the same place, minus the semicolon, it would result in awkward construction. The introduction of this phrase, unlike the converse case in English, automatically splits the sentence in two, or in two independent clauses which need to be divided by a semicolon.

resulting from (line 12) - Again, as in the case of *complications,* this also means *owing to*, so it is also translated as *debido a.*

extent (line 15) - This word has many possible translations. We chose *gravedad* because we felt it indicated the author's meaning in best way. In this case, since *extent* refers to an illness, and not a rash, for example, it refers to degree, not area or size. Degree, in this instance, refers to degree of severity, whether due to progression or prognosis. In any case, *gravedad* covers these possibilities quite well. This demonstrates how you should not be too literal, but should think carefully about what the author is really getting at, and thus not stray from the essence of the text.

Cartera Valor (Value Portfolio)

Crea riqueza a largo plazo

El objetivo principal de nuestra Cartera Valor es crear riqueza a largo plazo con riesgo moderado. Con un rendimiento anual compuesto del 18,6% (neto de cargos) desde su comienzo en 1974, esta cartera tiene un historial comprobado de satisfacer los objetivos de los clientes y representa el éxito más grande de Compu-Val.

Al seleccionar acciones para nuestra Cartera Valor, seguimos un principio de inversión conservador mediante la compra de acciones de bajo costo relacionado con el valor de los activos que controla la compañía. Para hacer esto, calculamos la relación **Valor/Costo** para cada compañía mediante la determinación del valor de mercado de la deuda más el capital social componente de Valor—y comparándolo al activo neto de la compañía ajustado por inflación—el componente de Costo.

Además, sólo compramos acciones de compañías lucrativas cuyo desempeño empresarial está en aumento así como se mide por el rendimiento de flujo de fondos de la inversión (cash flow return on investments, [CFROI]). El enfoque de CFROI elimina las distorsiones de contabilidad y da una medida verdadera del desempeño empresarial "ajustado a la inflación". Las comparaciones exactas entre compañías que podemos realizar llevan a un desempeño excepcional de las acciones.

Cuando evaluamos la compra de una acción, buscamos empresas que combinan deudas bajas y solidez financiera. Favorecemos aquellas compañías con productos o servicios de alta calidad, particularmente en mercados exclusivos que ofrecen oportunidades atractivas a largo plazo. Además, compramos compañías que crecen a una velocidad sostenida y enfocamos nuestra atención en los negocios núcleos.

Los valores de cartera son monitoreados diariamente y revisados cada semana por los profesionales de inversión de Compu-Val. Cuando el precio corriente de una acción alcanza nuestra meta, la vendemos. Cuando una acción ha perdido un 15% o más en comparación al costo promedio, es examinada por nuestro comité de inversión.

Cartera de Ingreso Fijo

Minimiza los riesgos para inversores conservadores

Nuestra Cartera de Ingreso Fijo fue establecida en 1984 para brindar una opción atractiva para aquellos inversores más conservadores y adversos al riesgo. La Cartera está diseñada para obtener el rendimiento máximo consistente con fluctuaciones limitadas de capital y un riesgo mínimo de crédito. Con una tasa de rendimiento anual compuesto del 9,6% (neto de cargos) desde su comienzo, nuestra Cartera de Ingreso Fijo ha dado resultados consistentes y positivos a los inversores cada año. Las inversiones están limitadas a bonos de gobierno e instrumentos de deuda corporativa clasificados "A" o mejor por los servicios de Standard & Poor/Moody de clasificación de tasa de bono. Los plazos de vencimiento son escalonados y la Cartera promedio madura en 6 años o menos. Las inversiones son seleccionadas según nuestro análisis de primas del riesgo comparado con los valores de gobierno sin riesgo.

value portfolio (line 3) - Phrases such as this occur more in the United States than anywhere else. These are fanciful terms which are created in the business, advertising, and media industries, many of which seem commonplace in everyday life in this country, but which, in practical terms and at first sight, seem impossible to translate. The best way of handling this challenge is to try to analyze the author's meaning as best you can. This can be accomplished not by being too literal (which could result in an absolutely silly and meaningless phrase) but by thinking carefully about what the author is really getting at, and thus not straying from the essence of the text. In this instance *cartera valor* was the best approximation, which is sometimes all you can get, no matter how hard you try. That is the reason for putting its original name in parentheses, because it ***is*** a name, and since an approximation was as close as we could get, it is important to let the reader know this. (If the original name means nothing to them, it will not really matter, because chances are that the translated version did not mean anything to them either.) This is ***not*** to be interpreted as carte blanche to cover yourself by putting the original in parentheses whenever you are stuck (otherwise your business will dwindle quickly!), but in cases where the original term was a made-up, albeit proper, term, it is all right to do that to further inform the audience.

A similar case is presented by the example of *niche markets* (line 17), although this is not a proper name, but rather a "catch-all" phrase that has meaning in the United States, though maybe not in other countries, even if translated as perfectly as possible. So, once again, you need to examine the context and come up with the closest approximation of what the author is referring to, keeping the context in mind above all else.

creating wealth over ... (lines 5-6) - The sentence order was completely changed upon translation, yet the idea the translated version conveys is true to the original. The order chosen, however serves the target language best in that it flows better than if the order had been kept the same. There is never a need to be literal, as we have been pointing out repeatedly. Accuracy and good writing are not mutually exclusive, even in the world of translation.

compounded annual return (line 6) - This phrase, although perhaps coined artificially for purposes of public consumption does not have to analyzed in terms of the author's intention. A good commercial/business & finance dictionary can give you subject-specific words separately, if not in compounded (no pun intended) form, which will allow you to put together the best phrase you can come up with if no standard is available.

CFROI (line 15) - Since this refers to a subject-specific term, it does not make sense to translate it literally and make up a new acronym for it. Instead, it should be translated first (with the original in brackets to let the audience know that it "came from" somewhere else (i.e., you did not make it up) and then give the real acronym that will be used throughout.

maturities (page 2, line 2) or *plazos de vencimiento*. This is where you can clearly see the need to keep up with the times. The Spanish you speak daily, in addition to not being necessarily technical in any way, may also be dated. The same occurs in English. New terms are created constantly in a wide variety of fields, and even the best dictionaries may not be that up-to-date (nor do you want to buy a new one for the same field every year). The solution is to haunt the library, keep up with daily, weekly, or monthly subscriptions in the areas you claim as your areas of expertise, and use your personal contacts. Thorough research ***will*** put you ahead of the majority of translators, getting you noticed by your clients and rewarded.

Section 3

Where Do I Go from Here?

After you have completed all the assignments, or at least those you have chosen to do, you have to make a major decision. If you have never translated for profit before, you have to ask yourself if you are ready to try your hand at soliciting for-profit translation work. If the answer is yes, contact one or two of the enclosed work sources (Appendix E) and apply for a test or a short assignment. If you have translated for profit before in a specific area, such as legal documents, and you feel you are now ready to venture into another area, such as medical translation, first try out some short assignments in that field, until you become comfortable with it.

It is important that you develop a relationship with one or more translation companies in your own geographical area. There is absolutely no reason in this day and age for a translator, say, in Wisconsin, not to develop a profitable relationship with a translation company in Florida, and in time you may indeed find out that a certain company on the other side of the continent is your best source of work. But as a beginner, there are many advantages to interfacing personally with the staff of a translation company close to home. They can give you a great deal of practical advice, help you with the hardware and the software, and share other kinds of experience with you. There is an old saying: "A good neighbor is better than a distant relative."

If, on the other hand, you feel you are not ready to take the plunge, you may want to consider a translation course or study program. Refer to Appendix 4 in the book, and see if there is a program close to home you can sign up for. Or you may decide to continue practicing on your own.

Whatever you decide to do, don't give up too easily. If you do enjoy the challenge of translation, continue to improve you skills. Remember: Translation can be a full-time career or a very fulfilling and profitable sideline. It is definitely the kind of expertise worth having, no matter what else you choose to do with your time.

Appendices

Appendix A. Requisites for Professional Translators

Any person who knows more than one language has the ability to explain a word or a sentence in what translators call "the source language" (the language you translate from) by using an equivalent word or sentence in what they call "the target language" (the language you translate into). This, in effect, is the beginning of translation. But it is only the beginning. It does not automatically turn a person into an accomplished translator. Along with the knowledge of the source and the target language, a translator must have an aptitude for translation. Some people are endowed with a talent for translation. It is not an acquired skill, like riding a bicycle. It is rather a talent, like playing the violin. Some people have it and some don't. It is not necessarily an indication of a lower or higher IQ. Nor is it an indication of how linguistically gifted one is. It is an inborn skill that enables a person to change a text from one language into another quickly and accurately, or, if you will, think in more than one language at the same time. If you possess this skill, then it behooves you to develop it and make use of it, because there is never an overabundance of good translators, and it is almost axiomatic that the good ones can always find either full-time or part-time work.

The **first** requisite for the working translator is a thorough knowledge of both the source and the target languages. There is no point in billing oneself as a translator if one is not fully familiar with both languages, or does not possess a vocabulary in both equal to that of a speaker of those languages who has a university education or its equivalent.

The **second** requisite is thorough "at-homeness" in both cultures. A language is a living phenomenon. It does not exist apart from the culture where it is spoken and written. It communicates not only the names of objects and different kinds of action, but also feelings, attitudes, beliefs, and so on. To be fully familiar with a language, one must also be familiar with the culture in which the language is used, indeed, with the people who use it, their ways, manners, beliefs and all that goes into making a culture.

Third, one must keep up with the growth and change of the language, and be up-to-date in all of its nuances and neologisms. Languages are in a constant state of flux, and words change meaning from year to year. A pejorative term can become laudatory, and a neutral term can become loaded with meaning. Thirty years ago the English word "gay" simply meant "joyous." Now it is used to define an entire segment of society. We once spoke of the "almighty dollar." Now as we travel abroad we may find out the dollar is not necessarily everyone's preferred currency.

Fourth, a distinction must be made between the languages one translates from and into. Generally speaking, one translates from another language into one's own native language. This is because one is usually intimately familiar with one's own language, while even years of study and experience do not necessarily enable one to be completely at home with an acquired language. The exceptions to this rule are usually those people who have lived in more than one culture, and have spoken more than one language on a regular basis. Those may be able to translate in both directions. There are also rare gifted individuals who have mastered another language to such a degree that they can go both ways. They are indeed extremely rare. Given all of this, one should allow for the fact that while the ability of the accomplished translator to write and speak in the target language (i.e., one's native tongue) may be flawless, that person may not necessarily be able to write excellent prose or give great speeches in the source language (i.e., the language from which one translates). Then again, it is not necessary to be able to write and speak well in the language one translates from, while it is to be expected that a good translator is also a good writer and speaker in his or her native language.

Fifth, a professional translator has to be able to translate in more than one area of knowledge. Most professional translators are called upon to translate in a variety of fields. It is not uncommon for a translator to cover as many as twenty or thirty fields of knowledge in one year, including such areas as political subjects, economics, law, medicine, communications and so on. Obviously, it would be hard to find a translator who is an economist, a lawyer, a medical doctor, and an engineer all wrapped into one. In fact, such a person probably does not exist. One does not have to be a lawyer to translate legal documents. Many a professional translator has been able to gain enough knowledge and acquire a vocabulary in a variety of technical fields to be able to produce perfectly accurate and well-written translations in those fields. This is not nearly as difficult as it may seem, since most technical fields utilize a well-defined number of terms which keep repeating themselves, and as one keeps translating the same subject, they become more and more familiar to the translator. One must, however, have a natural curiosity about many different areas of human knowledge and activity, and an interest in increasing one's vocabulary in a variety of related as well as unrelated fields.

Sixth, an effective translator must have a facility for writing or speaking (depending on whether the method used is writing, speaking, or dictation), and the ability to articulate quickly and accurately, either orally or in writing. Like a reporter, a translator must be able to transmit ideas in real time, and in good understandable language. Translation is a form of writing and speech-making, and a translator is, in a sense, a writer and an orator.

Seventh, a professional translator must develop a good speed of translation. There are two reasons for this: First, most clients wait until the last minute to assign a translation job. As a result, they turn to a translator or a translation service with what is perhaps the most typical question in this business: "How soon can you have this job ready for me?" The professional translator has to be prepared to accept that long job with the short turnaround time, or there will be no repeat business from that particular client or from most other clients, for that matter. Secondly, translation is generally paid by the word. The more words one can translate per hour, the more income one will generate. Translating 50 words per hour can land a translator in the poorhouse. Serious translation starts at 250 words per hour, and can reach as high as 1000 words per hour using word processing, and close to 3000 words per hour using dictation (the author actually knows such a translator). High volume translators are the ones who will be the most successful.

Eighth, a translator must develop research skills, and be able to acquire reference sources which are essential for producing high quality translation. Without such sources even the best of translators cannot hope to be able to handle a large variety of subjects in many unrelated fields. Dedicated translators are the ones who are always on the lookout for new reference sources, and over time develop a data bank which can be used in their work.

Ninth, today's translator cannot be a stranger to hardware, software, fax, modem, the Internet, and the latest developments in all those media. Translation has become completely dependent on electronic tools. Gone are the days of handwriting, the typewriter, and all the other "prehistoric" means of communication. The more one becomes involved in translation, the more one finds oneself caught up in the latest high-tech developments.

Tenth, a translator who wishes to be busy on a fairly regular basis doing translation work must carefully consider the fact that certain languages are in high demand, say, in Washington or in Los Angeles, while others are not. Thus, for example, there is high demand for Japanese, German, Spanish, French, Chinese, Arabic, Russian and Italian in both Washington and Los Angeles, but not nearly as much for Bulgarian, Farsi, Czech, or Afrikaans. If your language falls within the second group, it is extremely advisable to also have language expertise in one of the languages of the first group, or to seriously consider whether your particular language has enough of a demand to warrant a major investment of time and effort on your part. One should always check and see what kind of a potential one's language specialty has in a given geographic area.

The above ten points are the essential criteria for developing a translation career. There are many other considerations, but none as important as these. If you feel that you can meet all of the above criteria, then you should continue reading this handbook and putting it to good use.

The Well-Rounded Translator

The main division in the translation field is between literary and technical translation. Literary translation, which covers such areas as fiction, poetry, drama, and the humanities in general, is often done by writers of the same genre who actually author works of the same kind in the target language, or at least by translators with the required literary aptitude. For practical reasons, this handbook will not cover literary translation, but will instead focus on the other major area of translation, namely, technical translation. High quality literary translation has always been the domain of the few, and is hardly lucrative (don't even think of doing literary translation if your motive is money), while technical translation is done by a much greater number of practitioners, and is an ever-growing and expanding field with excellent earning opportunities. This chapter discusses the characteristics of the well-rounded technical translator.

The term "technical" is extremely broad. In the translation business it covers much more than technical subjects in the narrow sense of the word. In fact, there is an overlap between literary and technical translation when it comes to such areas as social sciences, political subjects, and many others.

One way of defining technical translation is by asking the question, does the subject being translated require a specialized vocabulary, or is the language non-specialized? If the text being translated includes specialized terms in a given field, then the translation is technical.

The more areas (and languages) a translator can cover, the greater the opportunity for developing a successful translation career. Furthermore, as one becomes proficient in several areas, it becomes easier to add more. Besides, many technical areas are interrelated, and proficiency in one increases proficiency in another. In addition, every area breaks down into many subareas, each with its own vocabulary and its own linguistic idiosyncrasies. Thus, for example, translating in Arabic does not make one an expert in all spoken Arabic dialects, yet a knowledge of several of those dialects is very beneficial for the professional Arabic translator.

How does one become a well-rounded translator? The answer can be summed up in one word—experience. The key to effective translation is practice. Since human knowledge grows day by day, and since language keeps growing and changing, the well-rounded translator must keep in touch with knowledge and language on a regular basis. The worst thing that can happen to a translator is to be out of touch with the source language for more than a couple of years. What the rusty translator may find out is that new words, new concepts and new ways of using those words and applying those concepts have come into being during that period of "hibernation," and one's old expertise is no longer reliable.

Translation, therefore, is a commitment one makes not for a limited period of time, but rather long-term. It is to be assumed that anyone who becomes a translator is the kind of person who loves words and loves the challenge of using words effectively and correctly. Such a person will not become an occasional translator, but will make translation a lifelong practice.

Good and Bad Translation Habits

The accomplished translator can develop good as well as bad habits. Starting with the bad, we have already pointed out oneClosing touch with the source language for long periods of time. Another bad habit is taking illegitimate shortcuts while translating. There are several types of such shortcuts. The most typical is failing to look up a word one is really not sure how to translate. Being ninety percent sure of a word's meaning is not good enough in professional translation. If one is not sure of a word's meaning, even after all available means have been exhausted, then one must put in a translator's note to that effect, or make it known in some other way that there is a problem with translating that particular word. Anything less would be deceptive.

Another illegitimate shortcut is summarizing a paragraph instead of providing a *full* translation. There is such a thing as summary translation of a paragraph or a document. If a summary is called for, then this is precisely what the translator is expected to provide. But the most common form of translation is what=s known in the business as a verbatim translation, which is a full and complete rendition of the source text. When verbatim translation is ordered, anything less than a full translation is an illegitimate shortcut. Unfortunately, some translators tend to overlook this from time to time, especially when they undertake more work than they can accomplish by a given deadline, and decide to summarize rather than miss that deadline.

Perhaps the worst habit for a translator is to decide at a certain point in time that his or her knowledge of either the source or the target language is so good that it cannot possibly stand any improvement. The moment one stops growing linguistically, one is no longer on the cutting edge of one's profession. The good translator is a perennial language student, always eager and willing to learn more and to keep up with the latest.

As for good habits, the most important, perhaps, are the ones we obtain by reversing the above-mentioned bad habits. But there are many more. One excellent habit is to read professional literature in the field one will be called upon to translate in with reasonable frequency. One good example is *Scientific American*, which can help anyone who translates subjects of science and technology to learn the style or styles used in scientific writing. People who work in the field of translating business documents should definitely read business periodicals, not the least of which is *The Wall Street Journal*. One does not have to be a scientist to translate scientific articles, or have a business degree to translate business documents, but a general understanding of the subject goes a long way towards providing an accurate translation of the subject.

Another excellent habit is to translate not only for profit but also for enjoyment and experience. Most people, unfortunately, are not so taken with their daily work that they would want to continue doing it after hours for fun or practice. But an accomplished translator is someone who will on occasion translate simply for the sake of sharpening his or her skill, or accept a very small fee because of personal commitment to the subject matter, or because of a personal interest. This writer, for example, enjoys translating poetry because of the challenge of doing what is perhaps the most difficult type of translation, and, quite simply, because of the enjoyment of poetry.

Yet another good habit is always to be on the lookout for dictionaries. Many dictionaries are hard to find, and are available in few places. This writer in all his travels across the United States and abroad always stops in bookstores to look for dictionaries. One can also order dictionaries from bookstores and from publishers, but then one has to know what to order and from whom.

The last good habit I would like to mention is the practice of compiling word lists and building a reference library. Dictionaries do not have all the words and terms a translator needs, nor do they contain all the information which specialized references may have. There are aids for translators put out by certain organizations, and there is professional literature in every field. In recent years there has been a growing awareness of the need for terminology management, and with the constant advances in computer technology databases have been proliferating, making the work of the translator much easier than ever before. Good references are worth their weight in gold when they are needed for a specific translation, and over time the experienced translator develops an extensive library of glossaries which become essential for any translation assignment.

Appendix B. Translation Techniques

Preliminary Considerations

You are given a text to translate. Before you commit yourself to doing any work on it, you must ask yourself a few preliminary questions. They are:

1. Is the text legible?
2. Am I familiar enough with the subject to tackle it?
3. Do I have the linguistic resources (dictionaries, human contacts) to decipher unfamiliar words?
4. Is the text complete, or are there any missing parts?
5. Can I do it within the requested timeframe?
6. Do I have a good reason for doing it (doing it as a learning experience, or because you enjoy it, or to help a friend, or because you are properly compensated for doing it)?

Once you have answered all the above questions to your own satisfaction, you are ready to proceed with the translation.

Effective Approaches

There is no single effective approach to translation, and over time translators develop personal techniques which enhance the quality and the speed of their translation. No one set of rules applies equally to everyone, but there are certain methods and means of translation which can help almost any translator achieve greater accuracy and output. The following is a review of some of the key techniques which are becoming almost universal among professional translators.

The first and foremost question a translator must deal with today is what kind of equipment to use in the process of translating. In the days of the pen and the typewriter this question was much less crucial. Today, however, translation has become almost totally dependent on computers, for several good reasons: (a) Word processing allows far greater flexibility in producing text than any other contemporary means. The output of most translators has been tripled and quadrupled through the use of computers; (b) Computers allow text to be stored on a disk and reprinted or modified later on, a function which is invaluable in the translation field; (c) Clients nowadays are getting used to asking for translation on disk, since it allows them to edit, reprint, modify and enhance the physical appearance of a document; (d) If more than one translator is involved in a given translation project, the text from the various translators can be entered by an editor on one disk and equalized or manipulated as necessary, without having to redo any particular portion thereof.

In addition, it is becoming more common every day to use electronic means such as a modem, fax or e-mail to transmit and receive text. These tools are no longer a luxury. Their cost has been coming down, and more and more translators are acquiring them. Many people today are saying they cannot imagine how translators were ever able to manage without them. The answer is very simple: manage we did, but it took us ten days to two weeks to do what we can now receive, translate and deliver in two or three days.

The next question when approaching a translation assignment is: Am I qualified to do this particular translation? Only an honest answer will do. If one is not sure, then chances are one should not tackle that particular task. One must feel confident about a particular assignment if the results are to be satisfactory. The exception to this rule is a case where a client cannot find anyone else to do that particular job, and for some good reason is either willing to take a chance or to receive less than a complete and fully accurate rendition. In such a case it should be made clear between translator and client that the translation is not legally binding.

Once the commitment is made to proceed with the job, the translator will spend some time going over the entire document—even if it is book-length—and do a realistic assessment of the following points:

a. How long will it take to translate the document?
b. What reference tools are needed to get it done?
c. What kind of preliminary steps are needed prior to the actual work of translating?
d. What special problems are related to the document,such as legibility of blurred or poorly copied text or difficult handwriting?
e. Does the document contain text in a language or languages other than the main source language, and, if so, can the translator handle that language?

Regarding the question of time, one can do a quick estimate of the length of the document by averaging words per line, times lines per page, times number of pages. An experienced translator has a pretty good idea of the number of words per hour he or she can translate. This is an essential feature of undertaking a professional translation job, since most clients have tight deadlines and tend to give repeat business to those translators known for keeping to their deadlines.

As for reference tools, if, for example, one is given a document about telecommunications, one should make use of on'=s own resources in that field and/or borrow from other sources whatever one needs to accomplish the task.

Preliminary steps prior to actual translation can include a consultation with an expert in a specialized technical field regarding a difficult term, phrase, paragraph or concept which the translator does not feel comfortable with. Having access to such experts is one of the translator's most cherished assets. It can make all the difference in the world between a correct and effective translation and one that misses the main point of the entire text. Another preliminary step is a trip to the local, regional or even specialized library to do some research on the subject.

The problem of legibility should be identified *before* one begins the task, not after. Sometimes the problem may start in the middle of the document and be so severe as to render the translation of the first part useless. In that case, the translator may have wasted a great deal of time. Sometimes the problem is minor, and does not affect the overall outcome of the translation. In other cases, the client may decide to proceed with the translation and simply put the designation [illegible] (between brackets rather than parentheses) wherever a word or part of the text cannot be deciphered.

Unbeknownst to client and translator, when a translation job is first assigned, there may be portions of text inside the source document in a language other than the main language of the document. This can happen in commercial, scientific and scholarly documents. It even happens in Tolstoy's novel *War and Peace,* when the author starts using French instead of Russian. This too should be detected prior to commencing the translation work, and a decision has to be made as to: (a) Does that text need to be translated? (b) Can the translator handle it? (c) Is it necessary to assign it to another translator?

Once all this preliminary work has been done, one is ready to proceed with the actual translation work.

Depending on the particular text, one should either start translating at this point, or, in the case of a text containing highly specialized terminology which may send the translator on frequent trips to the dictionary, one should first go through the document and make a list of as many unknown or uncertain terms as possible, and then spend some time looking them up and making a word list. This technique saves a great deal of time, since once a list is completed it is much easier to sail through the text, and the time spent initially on making the list is very short compared to the time wasted on repeated interruptions to look up words. Moreover, by first mastering the more difficult terminology of the text, one gains a much better understanding of the subject and is certain to produce a better translation. From the very start, make it a habit to compile word lists and glossaries of subjectspecific terminologies, and keep it in a computer database program for future reference. In time, these lists will become your most valuable translation tool.

One should also follow good work habits. Some translators, particularly those engaged in freelance work, tend to overdo it, especially during their "busy season," when they can generate a large income during a relatively short period of time. They will go for twelve or more hours a day, and before they know it they will start complaining of stiffness in the neck and shoulders, blurred vision, and fatigue. One should not translate more than eight hours a day. Six is ideal. Eight is tolerable, provided one takes a few short ten to fifteen minute breaks. Ten is pushing it. Over ten is definitely hazardous to your health.

Before you get ready to submit your translation go over it again, using the following checklist:

Omissions—did you fail to translate any particular word or phrase, or even paragraph?
Format—does your format follow the original (breaking into paragraphs, for instance)?
Mistranslations—did you mistranslate any particular word?
Unknown words—were there words you were not able to translate which you would like to explore further?
Meaning—did you miss the meaning of any phrase or sentence?
Spelling—did you misspell any word which the spell-check function on your computer did not catch?
Grammar—did you make any grammatical mistakes?
Punctuation—did you mispunctuate or miss any punctuation marks?
Clarity—did you fail to clearly convey the meaning of any particular part of the text?
Consistency—did you call something by one name and then by another without any good reason?
"Sound-alike" words—did you mistranslate a word because it looks or sounds like the word in your target language but in reality has a different meaning?
Style—are you satisfied with the way your translation reflects the style of the original text (for example, the original is written in a clear, direct style, while the translation sounds more complex and indirect?)

This checklist is by no means exhaustive, but it does cover the main areas a translator must pay attention to.

As was already explained, your personal computer is your best friend when it comes to translating, editing, and producing a final copy. One can learn a few basic commands, say, in WordPerfect or Microsoft Word, and start using the computer. But there is much more to software than entering, deleting and inserting text. The better acquainted with software you become, the more it will help you with translation. Learn how to do columns and tables, how to use special technical and scientific symbols, do graphic functions, use the spell-check and the thesaurus, create data bases for glossaries and for your own administrative records, and you will tackle a great variety of technical text in many fields at a speed that will amaze you. Remember: speed in translation is the most important thing next to language proficiency. Without it you will not be profitable, and you will be overrun by the competition. With an established record of fast accurate translation you can write your own ticket.

After a few years of using the computer you may want to consider dictation. Personally, I prefer a mix of PC and dictation. When I have an unusually long job and not enough time to do it in, I may revert to dictation. Otherwise, I prefer word-processing. One could argue that by dictating one gets more done and earns more, but there are other things to consider, such as the cost of transcription, the need to edit transcription, and the better control one has over writing than speaking. Some of us are natural speakers; others are writers.

One continues to develop translation techniques over time. One of the most wonderful things about translation, in my opinion, is the fact that your mind is never idle, never in a rut, but rather always being challenged by new tasks, new subjects, new knowledge, and the need to keep up with new developments in language, with different fields of human knowledge, and with the events of the world. As a translator in the Washington area since the late seventies, I have found myself in the middle of world events, beginning with the peace treaty between Egypt and Israel in 1979, when I met Begin, Sadat and Carter, and, more recently, in my daily dealings with events in post-Cold War Eastern Europe, with a strife-torn Middle East, the famine in Somalia, the new North American Free Trade Agreement (NAFTA) between the United States, Canada and Mexico, and the growing involvement of the U.S. space program with the space programs of other nations. Very few people cover as broad an area as a translator. Every day we in the translation business find new challenges, and have to solve new problems. As a result, we are always developing new techniques and finding new answers.

APPENDIX C: TRANSLATOR'S SELF-EVALUATION

The following criteria were developed some years ago by a U.S. Government agency for determining the skill level of a potential translator whom that agency might have liked to hire. You may want to read this chapter carefully to try to make an honest determination as to where on this scale you find yourself at this time. If you are below Level 2+, you need to keep practicing. If you are at Level 3 or higher, you can start doing some professional translating. After Level 4 you are ready for some serious translating, and at Level 5 you can start making a living as a translator.

Translator Skill Levels

Level 0

No functional ability to translate the language. Consistently misunderstanding or cannot comprehend at all.

Level 0+

Can translate all or some place names (i.e., street or city designations), corporate names, numbers and isolated words and phrases,
often translating these inaccurately.

In rendering translations, writes using only memorized material and set expressions. Spelling and representation of symbols (letters, syllables, characters) are frequently incorrect.

Level 1

Sufficient skill to translate the simplest connected written material in a form equivalent to usual printing or typescript. Can translate either representations of familiar formulaic verbal exchanges or simple language containing only the highest-frequency grammatical patterns and vocabulary items, including cognates when appropriate. Translated texts include simple narratives of routine behavior; concrete descriptions of persons, places and things; and explanations of geography and government such as those simplified for tourists. Mistranslations common.

In rendering translations, writes in simple sentences (or clauses), making continual errors in spelling, grammar and punctuation, but translation can be read and understood by a native reader used to dealing with foreigners attempting to translate his/her language.

Level 1+

Sufficient skill to translate simple discourse for informative social purposes in printed form. Can translate material such as announcements or public events, popular advertising notes containing biographical information or narration of events and straightforward newspaper headlines. Has some difficulty with the cohesive factors in discourse, such as matching pronouns with referents. In rendering translations, writing shows good control of elementary vocabulary and some control of basic syntactic patterns, but major errors still occur when expressing more complex thoughts. Dictionary usage may still yield incorrect vocabulary of forms, although can use a dictionary to advantage to translate simple ideas. Translations, though faulty, are comprehensible to native readers used to dealing with foreigners.

Level 2

Sufficient skill to translate simple authentic written material in a form equivalent to usual printing. Can translate uncomplicated, but authentic prose on familiar subjects that are normally present in a predictable sequence, which aids the translator in his/her work. Texts may include description and narration in context, such as news items describing frequently occurring events, simple biographical information, social notices, formatted business letters and simple technical material written for the general reader. The prose is predominantly in familiar sentence patterns. Some mistranslations.

In rendering translations, has written vocabulary sufficient to perform simple translations with some circumlocutions. Still makes common errors in spelling and punctuation, but shows some control of the most common formats and punctuation conventions. Good control of morphology of language (in inflected languages) and of the most frequently used syntactic structures. Elementary constructions are usually handled quite accurately, and translations are understandable to a native reader *not* used to reading the translations of foreigners.

Level 2+

Sufficient skill to translate most factual material in nontechnical prose as well as some discussions on concrete topics related to special professional interests. Has begun to make sensible guesses about unfamiliar words by using linguistic context and prior knowledge. May react personally to material, but does not yet detect subjective attitudes, values or judgments in the material to be translated.

In rendering translations, often shows surprising fluency and ease of expression, but under time constraints and pressure language may be inaccurate and/or incomprehensible. Generally strong in either grammar or vocabulary, but not in both. Weaknesses or unevenness in one of the foregoing or in spelling results in occasional mistranslations. Areas of weakness range from simple constructions, such as plurals, articles, prepositions and negatives, to more complex structures, word order and relative clauses. Normally controls general vocabulary, with some misuse of everyday vocabulary still evident. Shows a limited ability to use circumlocutions. Uses dictionary to advantage to supply unknown words. Translations are understandable to native readers not used to dealing with foreigner=s attempts to translate the language, though style is obviously foreign.

Level 3

Able to translate authentic prose on unfamiliar subjects. Translating ability is not dependent on subject matter knowledge. Texts will include news stories similar to wire service reports, routine correspondence, general reports and technical material in his/her professional field, all of which include hypothesis, argumentation and supported opinions. Such texts typically include grammatical patterns and vocabulary ordinarily encountered in professional reading. Mistranslations rare. Almost always able to correctly translate material, relate ideas and make inferences. Rarely has to pause over or reread general vocabulary. However, may experience some difficulty with unusually complex structures and low-frequency idioms.

In preparing translations, control of structure, spelling, and general vocabulary is adequate to convey his/her message accurately, but style may be obviously foreign. Errors virtually never interfere with comprehension and rarely disturb the native reader. Punctuation generally controlled. Employs a full range of structures. Control of grammar good, with only sporadic errors in basic structures, occasional errors in the most complex frequent structures and somewhat more frequent errors in low-frequency complex structures. Consistent control of compound and complex sentences. Relationship of ideas presented in original material is consistently clear.

Level 3+

Increased ability to translate a variety of styles and forms of language pertinent to professional needs. Rarely mistranslates such texts or rarely experiences difficulty relating ideas or making inferences. Ability to comp-

prehend many sociolinguistic and cultural references. However, may miss some nuances and subtleties. Increased ability to translate unusually complex structures and low-frequency idioms; however, accuracy is not complete.

In rendering translations, able to write the language in a few prose styles pertinent to professional/ educational needs. Not always able to tailor language to suit original material. Weaknesses may lie in poor control of low-frequency, complex structures, vocabulary or the ability to express subtleties and nuances.

Level 4

Able to translate fluently and accurately all styles and forms of the language pertinent to professional needs. Can translate more difficult prose and follow unpredictable turns of thought readily in any area directed to the general reader and all materials in his/her own special field, including official and professional documents and correspondence. Able to translate precise and extensive vocabulary, including nuances and subtleties, and recognize all professionally relevant vocabulary known to the educated nonprofessional native, although may have some difficulty with slang. Can translate reasonably legible handwriting without difficulty. Understands almost all sociolinguistic and cultural references.

In rendering translations, able to write the language precisely and accurately in a variety of prose styles pertinent to professional/ educational needs. Errors of grammar are rare, including those in low-frequency complex structures. Consistently able to tailor language to suit material and able to express subtleties and nuances.

Level 4+

Increased ability to translate extremely difficult or abstract prose. Increased ability to translate a variety of vocabulary, idioms, colloquialisms and slang. Strong sensitivity to sociolinguistic and cultural references. Increased ability to translate less than fully legible handwriting. Accuracy is close to that of an educated translator, but still not equivalent.

In rendering translations, able to write the language precisely and accurately, in a wide variety of prose styles pertinent to professional/educational needs.

Level 5

Can translate extremely difficult and abstract prose (i.e., legal, technical), as well as highly colloquial writings and the literary forms of the language. Translates a wide variety of vocabulary and idioms, colloquialisms, slang and pertinent cultural references. With varying degrees of difficulty, can translate all kinds of handwritten documents. Able to understand how natives think as they produce a text. Accuracy is equivalent to that of a well-educated translator.

In rendering translations, has writing proficiency equal to that of a well-educated native. Without nonnative errors of structure, spelling, style or vocabulary, can translate both formal and informal correspondence, official reports and documents and professional/educational articles, including writing for special purposes which might include legal, technical, educational, literary and colloquial writing.

APPENDIX D: SPANISH TECHNICAL DICTIONARIES

General Dictionaries

Diccionario de la lengua española, 22st ed., Madrid: Real Academia de la Lengua Española 2001 *Spanish-Spanish (Available in one volume hardcover deluxe edition, also 2-volume economic hardcover; CD-ROM edition forthcoming). This is the official dictionary of the Spanish language. The new 22nd edition incorporates for the first time foreignisms (mainly English) as well as many Latin American colloquialisms, making this dictionary a must for Spanish translators.*

HarperCollins Spanish-English/English-Spanish Dictionary, Bradley, D., New York: HarperCollins 1997 *Contains 230,000 references and 444,000 translations.*

The Oxford Spanish Dictionary, Jaman, B.G., and Russell, R., Oxford: Oxford University Press 1997 Thumb indexed. *Spanish- English/English-Spanish (also on CD-ROM) Very good.*

Diccionario internacional Simon and Schuster, de Gamez, T., New York: Simon & Schuster 1997 *English-Spanish/Spanish-English*
Contains many Latin Americanisms.

Multicultural Spanish Dictionary, Martínez, A., ed., Rockville, MD: Schreiber Publishing 2002 *Everyday Spanish words as they differ from country to country*

Libro de Estilo "El País," 14th ed., 1998

The Interpreter's Companion (Spanish-English/English-Spanish), Mikkelson, H., Acebo: 1996 *Compilation of terms not found in other dictionaries, covering legal, medical, drugs, weapons, profanity and slang terms.*

Diccionario Ideas Afines, Corripio, F., 1994 *Recommended*

Collins Spanish-English English-Spanish Dictionary, Unabridged, 3rd ed., Smith, C., New York: Harper-Collins 1993 *Widely considered one of the best of its kind.*

Visual Dictionary, Corbeil, J.C., New York: Facts on File 1992

Qué es qué (What's What), Enciclopedia visual bilingüe, Maplewood, NJ: Hammond 1988 *A pictorial dictionary.*

The Collins Spanish Dictionary, Barcelona: Ediciones Grijalbo 1988 *Good for Spanish for Spain.*

Diccionario de dudas y dificultades de la lengua española, Seco, M., Madrid: Espasa-Calpe 1986

Diccionario de uso del español, 2 vols., Moliner, M., Madrid: Editorial Gredos 1980, 1987 *Spanish-Spanish. Excellent; also available on CD-ROM.*

Gran diccionario español-inglés, Garcia Pelayo y Gross, R., Ediciones Larousse 1983 *Very comprehensive. European-oriented.*

2001 Spanish and English Idioms, Savaiano,E., & Winget, L., Woodbur-y, NY: Barron 1976 *Not exhaustive, but quite useful.*

Diccionario de los idiomas inglés y español, Velázquez, M., Englewood Cliffs, NJ: Prentice-Hall 1972 *Spanish-English/English-Spanish Old, but still useful.*

Sinónimos castellanos, Garcia, R. & Sopena, R., Buenos Aires 1967 *Spanish-Spanish.*

Diccionario español de sinónimos y antónimos, Sainz de R., F.C. *Spanish-Spanish* 1995

Buenas y malas palabras, Orellana, M. Chile: Editorial Universitaria 1998

General Regional Spanish Dictionaries

Mexican Spanish, London: Lexus 1996

In progress: Cuban and Peruvian Spanish dictionaries by Anthony T. Rivas (Ref. ATA).

The Dictionary of Chicano Spanish, Galvan, R.A., Lincolnwood, IL: NTC 1995

La Traducción del Inglés al Castellano, Orellana, M., 1994

Nuevo Diccionario Lunfardo, Gobello, J., 1994 *Colloquial language of Buenos Aires by a noted author on the subject.*

Lexicón de colombianismos , 2 vols., Banco de la República, Bogota: Biblioteca Luis-Angel Arango 1983

Diccionario de venezolanismos, vol. 1, A-I, Academia Venezolana de la Lengua, Universidad Central de Venezuela, Facultad de Humanidades y Educación, Caracas: Instituto de Filología Andrés Bello 1983

Diccionario de americanismos, Steel, B.

Diccionario de mejicanismos, Santamaria, F., Mexico City: Editorial Porrua 1974 *Very useful for any Mexico-related translation.*

Spanish: Aviation/Aeronautics
Diccionario aeronáutico civil y militar inglés- español, Velasco, Madrid: Paraninfo 1995

Spanish: Business and Related Fields
Spanish Business Dictionary, Sofer, M., Rockville MD: Schreiber Publishing 2001 *Multicultural business Spanish with varying terms from major Spanish-speaking countries such as Spain, Mexico, Venezuela, Chile and Argentina.*
Spanish Dictionary of Business, Commerce and Finance, London: New York, Routledge 1997 ; CD-ROM 1998
Nuevo diccionario bilingüe de economía y empresa, Lozano, J.M., Madrid: Ediciones Piramide 1996 *English-Spanish/Spanish-English*
Dictionary of Business, Collin, P.H., Middlesex, UK 1993
English-Spanish/Spanish-English
English-Spanish Banking Dictionary, Esteban, R.G., Madrid: Editorial Paraninfo 1993
Diccionario Enciclopédico Profesional de Finanzas y Banca (3 vols.), 1992
World Bank Glossary (Spanish-English/English-Spanish), 1992
Dictionary of Accounting (Spanish-English with S-E/E-S vocabulary), Kohler, E., 1990
Harrap's Glossary of Spanish Commercial and Industrial Terms, Rodrigues, L., London: Harrap 1990 *English-Spanish/ Spanish-English*
Diccionario comercial inglés-español/español-inglés, Giraud, A., Barcelona: Editorial Juventud 1990
Dictionary of Modern Business, Robb, L., Washington D.C.: Anderson Kramer Associates 1960 *English-Spanish/ Spanish-English*

Spanish: Chemistry
Hawley - Diccionario de química y de productos químicos, Sax, I., Barcelona: Ediciones Omega 1987 *Best of its kind.*

Spanish: Communications
Spanish Dictionary of Telecommunications, London/New York: Routledge 1998 *Spanish-English/English-Spanish*
Diccionario del video inglés-español, Perales, Madrid: Paraninfo
Diccionario para Electrónica, Telecomunicaciones e Informática, Buenos Aires: EMEDE, S.A. 1986
Diccionario Terminológico de los medios de comunicación, Prieto, F., Marid: Ediciones Piramide *English-Spanish*

Spanish: Computers
Dictionary of Information Technology (Spanish-English/English-Spanish), Vollnhals, O., 1997 *Very good.*
English-Spanish, Spanish-English Dictionary of Electrical and Computer Engineering, Kaplan, S., New York: John Wiley 2001
Computer Dictionary (English-Spanish), Freedman, C., 1994
Diccionario comentado de terminología informática, Aguado, G., Madrid: Editorial Paraninfo 1994
Diccionario de informática inglés-español/español-inglés, Madrid: Ediciones Diaz de Santos 1993
Diccionario de informática inglés-español, Olivetti, Barcelona: Editorial Paraninfo 1993 *Rather small and limited.*
Dictionary of Computer Terms, Chiri, A., New York: Hippocrene 1993
Diccionario de microinformática inglés-español, Tapias, R., Barcelona: Editorial Noray 1985 *Old but useful.*

Spanish: Electronics
Diccionario de electrónica español-inglés, Amós, Madrid: Paraninfo
Diccionario de electrónica y técnica nuclear, Markus, J. *Spanish-English*

Spanish: Engineering
Electrical and Computer Engineering Dictionary, Kaplan, S., New York: John Wiley & Sons 1996 *English-Spanish/Spanish-English*
Diccionario técnico inglés-español, Malgorn, G., Madrid: Paraninfo 1991
Dictionary of Environmental Engineering, J. Villate, Miami: Edicion-es Universal 1979
Engineers' Dictionary, Robb, L., New York: 1976

English-Spanish/Spanish-English; still considered one of the best.

Spanish: Environment
Spanish Dictionary of Environmental Technology, London/New York: Routledge 1998 *Spanish-English/English-Spanish*
Spanish-English Dictionary of Environmental Science and Engineer-ing, Headworth, H. New York: John Wiley 1997

Spanish: Law
English-Spanish, Spanish-English Legal Dictionary, Kaplan, S., New York: John Wiley 1997
Legal Dictionary (Spanish-English/English-Spanish), Alcarez Varo, E., 1997 *Highly recommended.*
Bilingual Dictionary of Criminal Justice Terms, Benaman, Virginia, Connolly, Norma, & Loos, Scott., Longwood, FL: Gould Publications, Inc. 1996 *English-Spanish Small but good.*
Diccionario de Derecho, Duran, R., 1995
El Inglés Jurídico, Alcarez Varo, E., 1995
Diccionario Jurídico, Moro, T., 1994
Diccionario Jurídico Español-Inglés, Cabanellas de las Cuevas, G. & Hoague, E.C., Austin, TX: Butterworth Legal Publishers 1991 *Spanish-English/English-Spanish Highly recommended; complements the Alcarez Varo (see above).*
Dictionary of Law, Economics and Politics, Ramon Lacasa Navarro Madrid: Editoriales de Derecho Reunidas 1989
Diccionario de Derecho, de Pina,R. & de Pina Vara,R.,
Mexico City: Editorial Porrua 1985
Diccionario de términos legales, Robb, L., Mexico: Editorial Limusa 1980
Diccionario de ciencias jurídicas, políticas y sociales, Ossorio, M.

Spanish: Maritime
Nautical Dictionary (Spanish-English/English-Spanish), Malagón Ortuondo, J.M., 1996
Diccionario marítimo y de construcción naval, Perez, J.A., Barcelona: Ediciones Garriga 1976

Spanish: Medicine
Medical Dictionary: English, Spanish, Portuguese, Nolte-Schlegel, New York: Springer-Verlag 2001
English-Spanish/Spanish-English Medical Dictionary, Rogers, G. New York: McGraw-Hill 1997
The Delmar English-Spanish Dictionary for Health Professionals, Kelz, R.K., Delmar Publishers 1996
Medical Dictionary (Spanish-English/English-Spanish), McElroy, O.H. et al, 1996 *Good, small paperback.*
Mosby's Medical Dictionary (Spanish-English with E-S/S-E Vocabulary), De Teran Bleiberg, E., 1995 *Extensive explanations and illustrations.*
Bilingual Glossary for Medical and Healthcare Translators: Oncology, Hematology, and Radiotherapy: English-Spanish-English, Albin, V., and Coggins, M., Houston: PCM Translation Resources 1994
Medical Encyclopedic Dictionary (Spanish-English), 2 vols., Dorland 1992 *Excellent*
Diccionario de términos médicos, Torres, R., Madrid: Alhambra 1989 *Spanish-English/English-Spanish*
Diccionario de los Términos Técnicos de Medicina, Garnier, M., and Delamare, V., 1981

Spanish: Military
Diccionario técnico militar, Gomez, A., Madrid: Ediciones Agullo 1980
Diccionario moderno de tecnología militar, Wells, R., Fairfax, VA: Lexicon Press 1977

Spanish: Mining
Glossary of Mining Terms (Spanish-English/English-Spanish), Diaz Preto, P., 1995

Spanish: Psychiatry
Glossary of Psychiatric Terms (Spanish-English/English-Spanish), Nemiah, J.C, 1996
Dictionary of Psychology and Psychiatry (Spanish-English/English-Spanish), Kaplan, S., 1995

Spanish: Real Estate

Dictionary of Real Estate Business, Salles, M. Dearborn: 1997

Spanish: Technical

Polytechnic Dictionary of Spanish and English Languages, rev. ed., Beigbeder Atienza, F. 1997 *Highly recommended.*

Routledge Spanish Technical Dictionary, 2 vols., London/New York: Routledge 1996. Vol. 1 (Spanish-English); vol. 2 (English-Spanish). CD-ROM 1998

The Contractors Dictionary of Equipment, Tools and Techniques, Kennedy, F. New York: John Wiley 1996

Technical Glossary for Bilingual Technical Writers and Translators: English-Spanish/Spanish-English, Rodríguez, H., published by author, 1996

Diccionario técnico, Rafael Garcia Diaz, Mexico: LIMUSA, S.A. 1996

Glosario Internacional para el Traductor (Spanish-English/English-Spanish), (reprint) Orellana, M., 1998 *Very useful for translators*

Diccionario de términos científicos y técnicos, 5 vols., Barcelona: McGraw-Hill Boixarev 1981 *A translation of the excellent McGraw-Hill* ***Dictionary of Science and Technology*** *(see General References).*

Spanish-English/English-Spanish Encyclopedic Dictionary of Technical Terms, 3 vols., Collazo, J.L., McGraw-Hill 1980

Glosario español-inglés de términos técnicos, 5 vols., Thomann, A. E., Armco Steel Corporation 1975 *Spanish-English/English-Spanish*

APPENDIX E: SOURCES OF TRANSLATION WORK

> This book includes a CD (on the inside of the front cover) with lists of translations companies and international corporations who do a large volume of translation in all languages.

If you choose freelance translation, you should consider yourself a one-person translation company. Your main concern will be where to find work. The need for freelance translation is greater than anyone can estimate, and is clearly growing at a rapid rate. Worldwide, translation is a multibillion dollar industry. But finding translation work on your own is easier said than done. The main problem is that translation is hardly ever a steady, ongoing function of any particular work source, such as an embassy, a company, a government agency, or even a publisher. None of those needs translation every day of the year. Each of them may need a great deal of translation all at once (more than any one person can handle within the given time-frame), and then none for a long time. And, if any one of them needs translation on an ongoing basis, chances are a decision will be made to hire an in-house translator rather than farm out the work.

The fact remains, however, that a well-rounded freelancer can earn well over $50,000 a year, and, in the case of highly specialized technical translators in major languages like Spanish, German, Japanese or Russian, even $100,000 or more. The secret to all of this is establishing for yourself a good clientele. There are two ways of doing this. The first, and by far the hardest, is finding your own clients and working with them directly. You may want to contact embassies, law firms, publishers, government agencies and so on, and solicit work directly from them. If you are fortunate enough to find some good steady clients on your own, you will be doing quite well. But the problem often lies in the word "steady." What seems to be a steady client today may not be so steady tomorrow.

This brings us to the second, and by far the safer option, which is translation agencies. There are hundreds of them in the United States, and they handle huge amounts of translation business every year. In this chapter we will discuss translation agencies, as well as direct sources of translation available to the freelancer.

Translation Agencies

Translation agencies, also known as translation companies, or translation bureaus, are for the most part privately owned commercial establishments ranging in size from one or two employees to ten or more, but hardly ever larger than ten. Some are divisions of larger companies, such as Berlitz, which is primarily a language school and publisher, offering translation as a secondary function. Some specialize in one language only, such as Spanish, German, or Japanese. Most offer several languages, and quite a few bill themselves as offering "all languages." This last type is somewhat pretentious, since there are more languages in the world than any one person can identify. But what they really mean is that they will make the effort to find a translator in almost any language they may be called upon to translate.

As a general rule, translation agencies employ relatively few in-house translators, since the flow of work in any given language is usually uneven. Instead, they rely on the services of a network of hundreds of freelancers who can handle a great variety of subjects. Those freelancers are located all over the United States and even abroad. The ones who are most reliable and professional get the major share of the work, and some of them earn the above-quoted figures.

As a freelancer, you need to cultivate at least one such agency, preferably two or three. The problem in working with only one is that, with few exceptions, there may not be a steady flow of work coming out of any given agency in any given language, in subjects you are equipped to handle. Two or three will give you better coverage, and assure a better flow. On the other hand, you may find yourself in a situation where all three ask you to do something at the same time, and you may not be able to do it. You need to establish an understanding with your agencies that would make an allowance for such a scenario, so that you don't spoil your relationship with any one of them.

The worst thing you can do as a freelancer working with translation agencies is to overcommit yourself. Your most important personal asset is your reliability. Once you fail to meet deadlines (keep in mind—the agency stands to lose a client if deadlines are not met), your reliability becomes questionable, and if you do it once too often, you may soon find out that those phone calls from the agency offering you work assignments stop coming.

Where do you look for translation agencies? Below you will find a listing of hundreds of such agencies. You can find more in the Yellow Pages, or through the ATA (American Translators Association), which has local chapters around the country. My suggestion is to start with those close to home. In this day and age of international electronic communications, distance has little meaning. But then again, close to home still works, because you can meet the people there, befriend them, and in some instances even avail yourself of their dictionaries and other resources.

Keep in mind that a translation company has overhead, and also needs to make some profit to stay in business. They do the hard work of finding translation assignments, and therefore share with you the profit from the job. You can usually make more money by going directly to the client, and if you have enough of your own clients you don't need a translation company to send work your way. But most freelance translators do need those companies, which invariably provide a more steady flow of work than what a freelancer can get on his or her own.

The two things all translation companies appreciate and reward in a freelancer are honesty and loyalty. If you agree to a deadline, stick to it. Don't renege on it at the last minute. That's a sure prescription to spoil your association with your company. Equally important is not to go behind the company's back and try to solicit its own clients directly. Some companies will make you sign an agreement to this effect. Others will rely on the honor system. Don't abuse their trust. It usually doesn't pay off.

Direct Sources of Work

Working with translation agencies usually does not stop you from finding your own clients, as long as there is no conflict of interest with the agency's clients. It would be impossible to list here all the potential sources of direct translation work, since they include practically the entire human race (everyone needs a document translated at some point). But there are some major sources which ought to be mentioned, and here are some of the more important ones.

Law Firms

Law firms are a major source of translation work. Some of the larger firms hire full-time translators or staff members who are bilingual, especially if they do business on a regular basis with a foreign entity. Most firms use translators on an as-needed basis. Legal translation is a specialized field in which you need to acquire experience working with legal documents. There are several legal specialties, such as patent law, international law, immigration, and so on. Each specialty has its own style and terminology, which a translator needs to become acquainted with. As a freelancer, adding legal translation to your list of specialties is an excellent idea. You will find out that your volume of translation will increase considerably by doing so.

Quite often, a law firm needs both document translation and interpretation. Keep in mind that interpretation is a discipline separate from translation, and that there is a big difference between one-on-

one consecutive interpreting and simultaneous conference interpreting. If asked by a law firm to do both text translation and interpreting, be sure to find out first exactly what the assignments consist of.

Industry

Corporations doing business in other countries have to deal with documents originated in the languages of those countries or English documents that need to be translated into those languages. Here again we find the two approaches of either hiring translators or farming out work to freelancers and to translation services (or a combination of both). This field is perhaps the fastest growing source of translation in the closing years of the twentieth century. More and more major American companies are turning to international business as a way to offset the decline of business at home and to gain a share of the world market. Their need for translation is growing every day, and even those who have in-house translators are finding themselves using freelancers because of their volume of translation work.

If you are fortunate enough to form a relationship with a major company doing business overseas, you may find yourself in the enviable position of dealing with a major, steady source of translation.

How does one get work with major corporations? If you have a special expertise in their field of work, say you are a telecommunications expert and you would like to translate for Sprint, find out who handles outside vendors or services, and give them a call or drop them a note. It also helps to know someone in the company, who can do some of the legwork for you and put you in touch with the right people. As a general rule, this is not easy to do. But persistence does pay off some of the time, and even if only a few respond, it is worth the effort.

The U.S. Government

For several decades following World War Two, translation in the federal government enjoyed a boom. The onset of the Cold War resulted in wide-scale translation activities on the part of the U.S. Department of Defense (DoD), all the branches of the service (particularly the Army, Navy, and Air Force), and the Central Intelligence Agency (CIA). In addition, such multilingual organizations as the Voice of America (which started broadcasting in 1942 in 40 languages), sprang into being. These and other organizations employed a host of translators, and farmed out millions of words every year to be translated. Looking back, those were the feast years of government translation. During the nineties, however, a major shift has been taking place. Since 1992, there has been a sharp decline in government translations, and those translation agencies and freelancers who were dependent on government work for their bread and butter, are hurting. This is not to say that the U.S. Government is no longer a source of employment for translators. There are still many opportunities for translation in the government, offered by such bodies as the Language Services division of the U.S. Department of State (for in-house as well as freelance translators and interpreters), the Library of Congress, the U.S. Patent and Trademark Office, and so on (see list of U.S. Government Agencies, Appendix 4).

My translation company, Schreiber Translations, Inc. (STI), started out in life as a full-service translation company for the federal government. This was in the early eighties, when the boom was still on. We provided translation from and into over 50 languages and dialects, as well as interpreting, transcription, voice-over, graphics, and editing. Luckily for us, we realized early on that, as the world was changing, we were better off not being locked into government work, and we were able to develop a lucrative practice in the nongovernmental sectors. Today, government work makes up less than 50% of our total translation work.

What does all of this mean to linguists seeking translation work in the closing years of the twentieth century? There is no one simple answer. The government has been known to swing like a fast-moving pendulum. This author believes that before long the federal government will be launching new major translation projects, in such sectors as world trade and international relations. American interaction with other cultures and languages is far from diminishing. If anything, it is increasing day by day. One good example is the American space program, conducted by the National Aeronautics and Space Administration (NASA). Back in the sixties, there were two major space programs in the world—the Soviet and the American. Today, many countries in Western Europe and Asia have an active space program, and many of them are involved in cooperative efforts.

NASA is actively involved in this new international space scene, and the need for translation in this particular organization should not be underestimated. But all of this, of course, is speculative. The federal government is making a historical effort to reduce its enormous deficit, and since the lawmakers of our land are not necessarily attuned to the critical importance of translation, translation may well be a casualty of the budget wars. Translators would be well advised not to write off doing translation for the government, since it is, and will continue to be, the single biggest employer around. But they should put at least as much, if not more, effort into finding translation work outside the government, where indeed the field is all but limitless.

What are the pros and cons of translating for the government? For many linguists over the years, the government has offered job security. I know some fine translators who have worked for the government for many years and had interesting and fulfilling careers. On the other hand, career translators in the government are not paid exceptionally high salaries. Quite a few freelance on the side in an effort to supplement their income. As for freelancers who contract with the government, here we have mixed results. Some have found themselves a cozy niche and are kept busy on a fairly regular basis. Others go through the "feast and famine" syndrome. Some have been frustrated by spending a long time hunting for government assignments, with little result. To freelance effectively for the government, one should, in most instances, work in one of the top five languages the U.S. Government is interested in (Spanish, Russian, German, French and Japanese). Any other language, even Chinese or Italian, may not offer enough of a flow. One should also make many contacts.

On the downside, government agencies, with few exceptions, are notorious for not paying on time. A few years ago, the government passed a "prompt payment" act, which mandates 30-day net payment, with interest accruing thereafter. The problem is, more than a few agencies take at least another 30 days to process an invoice, claiming that it takes that long to get your invoice "into the system." Others take even longer, and simply neglect to add interest. If you try to get it, you may soon find out you waste more time going after it than it's worth. In some extreme cases—which are by no means rare—it can take six to twelve months to get paid. Your invoice simply gets lost in the system. Your remedy is to contact your congressperson, who will call that particular government office to make sure you get paid. The problem is, those folks in the finance office often resent being put on the spot in this manner, and while they will expedite your payment, you may not get work next time. It all sounds shocking, but having been there, it is my civic duty to let you know.

Another drawback of government freelancing is the incredible amount of paperwork you are asked to fill out in doing some of those jobs. I have done five and ten dollar jobs that required three detailed forms filled out, plus an invoice in six copies, mailed to three different offices (finance, procurement, and contract). It took me longer to do the paperwork than to translate, and clearly the whole thing was not worth my while. If you do a fair amount of freelancing for the government, you learn to overlook these excesses. Otherwise, you may have to decide whether it's all worth it.

A third major drawback to doing government work is the fact that quite often the translation standards of a particular agency may be quite low. As a result, you run the risk of developing poor translation habits, lack of attention to details, lack of sufficient editing and proofing, and so on. What that means in a practical sense is that when a more demanding job comes along, especially in the private sector, you may find out your performance is not satisfactory, and you may lose your client.

Having said all this, it behooves me to add that during the last seventeen years I have interacted with hundreds of government employees and dozens of agencies, and many of those have been a delight to work with, and overall I have been very fortunate to have this unique opportunity to take part in many historical events, including the peace treaty between Israel and Egypt, the Gulf War, the disarmament agreement with the former Soviet Union, the Bosnian peace agreement, and many more. Thus, working with the government as a linguist is by no means a dull experience, and if you learn how to take some of the problems in your stride, you can do quite well.

State and Local Government

Government at all levels, from the municipal to the federal, needs translation. At the local level we find more and more city and county government translating their pamphlets, brochures and other documents into the languages of their immigrant populations, notably Asian languages and Spanish. In addition, local government has an ongoing need for interpreters, mostly for the court system, but also for social services, hospitals, and other local institutions. At this time of tight budgets, this may not be the most lucrative field, but it is definitely worth exploring, since it is local, and mostly uses local linguists.

Major Organizations

Among the largest organizations that use a great deal of translation one should mention the United Nations, The World Bank, the World Health Organization, the Organization of American States, to cite only some of the better-known ones here in the U.S. Surely there are more in Europe, such as the World Court, the European Community, NATO, and many more. Most of these organizations use in-house linguists, and do not farm out translation work if they can help it. But quite often they have more documents to translate than they can handle in-house, and they look for outside help.

Publishers

Book publishers use freelance translators in many different ways. This is not an easy field to break into, particularly for the beginner. Many publishers turn to academia for translators, and if you are in academia and can translate from or into another language in your field of expertise, there is a chance you can get the work. Others turn to established translators with name recognition. But it is certainly worth trying, to query publishers and find out if they need your services.

Software Localization Companies

A fast-growing area of translation is software localization, or translation of software-related text into other languages and adapting the text to the target culture. Software localization companies specialize in computer subjects and in English-into-foreign language translation. They employ in-house translators, but they also use freelancers from time to time. Their work encompasses everything from computer manuals to localization of web-sites.

Networking

An excellent source of work for freelance translators is personal contacts with other translators. The ATA's local chapters are one place where translators meet and get to know each other. The annual conference of the ATA holds a networking session, which is very valuable. One also meets translators through personal contacts in the translation field.

The latest way to meet translators is on the Internet. Entities such as the Foreign Language Forum on CompuServe bring together translators from all over the world. All you need to do is post a message on one of those programs, and before you know it you get a response from someone in your own town or half way around the world.

Keeping in touch with other translators is a prime means of finding out about work sources and assignments. There are clearly many advantages to networking in the translation field, and the more contacts one has the better.

The following is a list of translation companies who use freelance translators. The companies are not rated. It is up to you to find out which are best suited to your needs.

-A-

ABC Worldwide Translations & Interpretations, 8306 Wilshire Boulevard, Ste. 200, Beverly Hills, CA 90211
Phone: (310)260-7700 Fax: (310)260-7705
e-mail: info@wordexpress.com Website: www.wordexpress.net
Contact: Muriel Redoute Member: ATA

ABS Translation & Interpreting Service, 1685 Hampton Road, Abington, PA 19001 Phone: (215)886-2219 Fax: (215)886-7671
e-mail: ABStrans@aol.com
Contact: Aram Sarkisian Member: ATA

A.C.E. InfoTech Group, 1450 114th Ave SE, Ste. 1-230, Seattle, WA 98004
Phone: (425)453-4444 Fax: (425)453-0188
Website: www.aceinfotech.com
Contact: Maya Vengadasalam Member: ATA

AE Inc. Translations, 15995 N. Barkers Landing, Ste. 111, Houston, TX 77079 Phone: (281)870-0677 Fax: (281)556-9737
e-mail: translations@aetrans.com Website: www.aetrans.com
Contact: Stephen D. Ross, President Member: ATA

ABLE Innovations, Inc., 1100 Lakeway Drive, Ste. 200, Bellingha-m, WA 98229 Phone: (360)714-1390
e-mail: sales@ableinnovations.com
Website: www.ableinnovations.c-om Member: ATA

Academy of Languages Translation & Interpretation Services, 216 First Ave. South, Ste. 330, Seattle, WA 98104
Phone: (206)521-8601
Fax: (206)521-8605 e-mail: translate@aolti.com Website: www.aolti.com
Contact: Chaouky Kaboul Member: ATA

Accent on Language, 160 East 56th St., 5th Fl., New York, NY 10022 Phone: (212)486-6790 Fax: (212)486-2439
e-mail: info@accentonlanguage.com Website: www.accentonlanguage. com
Contact: Brian E. McDermott Member: ATA

ACCENT Translations, Inc.; 442 Beaulieu Road, Neenah, WI 54956; Phone: (920) 751-8260
Fax: (920) 969_0320; E_mail: accent@ameritech.net; Contact: Kimberlee A. Heier, President Member: ATA

Accento, The Language Company, 5527 Dyer St., Dallas, TX 75206
Phone: (214)363-5353 Fax: (214)363-0067

e-mail: accento@accento.net Website: www.accento.net
Contact: James Mahler Member: ATA

Access Transport & Translation Services, Inc., 528 S North Lake Boulevard, Ste. 1000, Altamonte Springs, FL 32701
Phone: (888)748-7575 Fax: (407)330-7959
e-mail: gengelman@accessontime.com Website: www.accessontime.com
Contact: Stacey L. Whidden Member: ATA

Accura International Translations, Ltd., 915 S. High St., Denver, CO 80209 Phone: (303)280-1292 Fax: (303)280-1489
e-mail: accura@ix.netcom.com Website: www.accuratrans.com
Contact: David Anderson Member: ATA

Accurapid Translation Services, Inc., 806 Main St., Poughkeepsie, NY 12603 Phone:(845)473-4550 Fax: (845)473-4554
Website: www.accurapid.com
Contact: Gabe Bokor
Founded 1978. Member: ATA. Translates German, Spanish, French, Hungarian and Japanese both from and into English, and Portuguese and Russian from English. Emphasis on Engineering, Patents, Law, Business and Finance trans-lation. Prospective translators should demonstrate competence and a professional attitude. Resumes are filed for future reference. The company pioneered the use of technological tools in the industry. The company counts freelance translators from Europe, Asia and the Americas in its professional pool. Maintains a pool of 600 translators.

Accurate Spanish Translations by Spanish Business Services, Inc., P.O. Box 720519, Dallas, TX 75372
Phone:(214)821-2050 Fax: (214)821-7065 e-mail: sbscorp@swbell.net Website: www.spanishbusiness.c-om
Contact: Robert H. Brandao Member: ATA

ACCUWORLD, LLC, 200 West Madison St., Ste. 930, Chicago, IL 60606 Phone: (312)641_0441 Fax:(312)641_7370
E_mail: ts@inlinguachi.com
Contact: Human Resources Department
Founded 1968. Member: ATA. Translates all languages, both from and into English. Covers all areas, with specialties in foreign language voice_overs, health care, pharmaceuticals, insurance, and technical manual translation. Resumes are compiled in a database according to area of expertise. The company, headquartered in Hartford, Connecticut has over 300 worldwide locations. It provides language training, cultural training, translation/interpretation services, and foreign language typesetting and voiceovers. Maintains a pool of 800+ translators.

Acentos, Marketing & Advertising & Translations, 400 E. Lake St., Minneapolis, MN 55408 Phone: (612)824-6109 Fax: (612)825-1729
e-mail: fcurbelo1@csc.om
Contact: Franklin Curbelo Member: ATA

AD-EX WORLDWIDE, 1733 Woodside Rd., Ste. 115, Redwood City, CA 94061 Phone: (800)223-7753 Fax: (650)474-2531
e-mail: translations@ad-ex.net Website:www.ad-ex.net
Contact: Robert Addis
Founded 1957. Member: ATA. Translates in all major languages, in all industrial fields, especially in the power and aerospace industries. Prospective translators must have demonstrable years of experience and written evidence of competence, including resumes and work samples. Resumes are responded to with a request for samples which are then thoroughly examined. The company is always glad to hear from veteran professional translators. Their client base is worldwide. Maintains a pool of "hundreds" of translators.

Adaptive Language Resources, Inc., 45 Mount Auburn St., Watertown, MA 02139 Phone: (617)924-0554 Fax:(617)924-0280

Website: www.adaptivelanguage.com
Contact: Adrian Spidle   Member: ATA

Advance Language Studios, 500 N. Michigan Ave., Chicago, IL 60611 Phone:(312)782-8123 Fax:(312)782-8356
Contact: Patrick Schoorlemmer
Founded 1963. Member: ATA. Translates from English into Spanish, French, German, Italian, Portuguese, Dutch, Japanese, Korean, Mandarin, Russian, and Polish. Emphasis on Technical, Advertising, Financial, Legal, and Commercial translations. Prospective translators should be born and raised in country of target language. Resumes are entered in a database. Company services include Translation, Interpreting, Desktop Publishing, Video dubbing in in-house studio. Maintains a pool of 1000+ translators.

Advanced Communication and Translation, Inc., 4332 Montgomery Avenue, Suite A, Bethesda, MD 20814 Tel: 301-654-2890 Fax: 301-654-2891; Web Page: http://www.act_translate.com
E-mail: act@act-translate.com
Contact: Monique-Paule Tubb   Member: ATA

Advanced Global Services, Inc., P.O. Box 965, Chesterton, IN 46304 Phone: (219)921-1733 Fax: (219)921-0640; E-mail: ags-inc@ags-inc.com Web site: www.AGS_INC.COM
Contact: Sylvie Prudhomme   Member: ATA

Advanced Language Translation Inc.; 25 N. Washington St, Rochester, NY 14614; Toll Free: 800-218-9024 Tel: 585-697-0462 X18 Fax: 585-697-0467 E-mail: info@advancedlanguage.com Web Site: www.advancedlanguage.c-om
Contact: Scott Bass, President Member: ATA

Affinity Language Services, 101 S. Fairview Ave., Wind Gap, PA 18091
Phone: (610)863-3955   Fax: (610)863-8571
e-mail: info@affinity-languages.com   Website: www.affinitylanguages.
com
Contact: Michelle Zuccarini
Founded 1998. Member: ATA. Works in Spanish, German, French, Italian, and Dutch. Also in Japanese. Main areas: Technical, sci-tech, legal, medical, and business. Also provides interpretation, editing, proofreading, transcription, and copywriting. Uses translators worldwide.

Agnew Tech-II, 741 Lakefield Rd., Ste. C, Westlake Village, CA 91361 Phone: (805)494-3999 Fax: (805)494-1749 e-mail: agnewi @agnew.com Website: agnew.com Contact: Irene Agnew
Founded 1986. Member: ATA, NAWBO, WITI. Translates Spanish, Chinese, French, German, Japanese, Russian, Vietnamese, etc. from English. Also scripts and voiceover. Covers all technical areas. Prospective translators should have a B.A. degree or higher and 2-3 years experience. Resumes received via e-mail, reviewed, graded and scheduled for a translation test. The company is a full-service translation bureau, as well as providing desktop publishing, web page design, multimedia, and audiovisual services. Maintains a pool of 300-400 translators.

AimTrans, Inc., 14752 Beach Blvd, Ste. 201, La Mirada, CA 90638 Phone: (714)522-8200 Fax: (714)522-7773 Website: www.aimtrans.com Contact: Irene Agnew   member: ATA

Airspeed Translations, 4200 Westheimer Road, Ste. 201, Houston, TX 77027
Phone: (713)586-2577 e-mail: rknauth@att.net
Website: www.airspeedtranslations.com
Contact: Rick Knauth

Alanguage Bank, Inc., 150 Lafayette St., 5th Floor, New York, NY 10013 Phone: (212)213-3336 Fax:(212)685-7263
e-mail: peiwens@alanguagebank.comWebsite: www.alanguagebank.com
Contact: Pei Wen Shih Member: ATA

Albanian Translation Services, 3738 Katie Place, Triangle, VA 22172
Home: (703)221-8837 Fax: (703)221-8836
E_mail: tirana@msn.com
Contact: Mirela M. Kuhn Member: ATA

Albors and Associates, Inc., P.O. Box 5516 Winter Park, FL 32793 Phone: (800)785-8634Fax: (407)657-7004
e-mail: rene@albors.com Website: www.albors.com
Contact: René A. Albors, President
Founded 1996. Member: ATA, NAJIT, Hispanic Chamber of Commerce, Orlando Chamber of Commerce. Translates Spanish, French, Portuguese, Japan-ese, German, Chinese, Russian, Polish, Arabic and Italian, both from and into English. Emphasis on Legal, Medical and Business translations. Prospective translators should be experienced, expert with terminology, and prompt in their delivery of finished work. Accepts unsolicited resumes for translation and inter-pretation. Resumes are computer-filed, and a letter of thanks is sent to acknowledge receipt. Maintains a pool of 5300 translators.

Alexandria Translations, 8827 Fort Hunt Rd., Alexandria, VA 22308 Phone:(703)799-7606 Fax:(703)799-7607 e-mail: 102554.3407 @compu-serve.com Contact: Lidia Terziotti Member: ATA
Translates and edits documents in a variety of languages.

Allegro Translations, 4513 Vernon Blvd., Madison, WI 53705
Phone: (608)233-3208 Fax: (608)233-3511 e-mail: info@allegro.com
Website: www.allegro-translation.com
Contact: Yaël Bratzlavsky
Founded 1993. Member: ATA. Translates English into Spanish, German, French, Japanese, Chinese, Russian, Italian, Brazilian Portuguese, Hmong and Laotian. Technical manuals, user manuals, employee handbooks, and marketing. Has over 100 translators.

Allen Translation Service, Box 1529, Morristown, NJ 07962
Phone: (973)292-2737 Fax: (973)292-3954 Contact: Jeanette C. Fredericks Member: ATA

Allslavic Translation Services; 4930 NW 84th Ave., Ft. Lauderdale, FL 33351
Phone: (800)775-5504 Fax: (954)741-3898
e-mail: info@slavprom.com Web site: www.slavprom.com
Contact: Stanka Moskov Member: ATA

ALLSPEAK Interpreting Services, P. 0. Box 1606, Glendale, CA 91209
Phone: (818)246-1515 Fax: (818)246-4211 e-mail: allspeakint@aol.com Contact: Tatiana E. Vorobieff Member: ATA

Allworld Language Consultants, Inc., P.O. Box 2128, Rockville, MD 20847
Phone: (301)881-8884 Fax: (301)881-6877 Website: www.alcinc.com
Contact: Carlos Scandiffio Member: ATA

Alpha Tech Communications, 4440 Chastant St, Suite E, Metairie, LA, 70002; Phone: (504)454_6554 Fax: (504)454_6717; Web Page: www.ustranslations.c-om E_mail: alphaTC@aol.com; Contact: Giovanna Icaza Director of International Language Services Member: ATA
Founded in 1998. Alpha Tech Communications translates many languages including English into Spanish, French, German, Portuguese, Chinese and Arabic. Along with translating English into those languages they translate them back into their respective language. They also translate Italian_English, Japanese_English, Greek_English, Russian_English, Etc. Their main subjects of translation are medical handbooks and legal documents. They deal with a wide variety of documents, from journal articles to

technical documents. They do accept resumes and they do use freelance. The resumes are reviewed, evaluated, classified and kept in their data files to contact freelancers when necessary. They require you to have a good command of the language pair(s) you work with. Experience, commitment and responsibility. Computer programs and internet tool skills. They have one in_house translator and about 20 freelancers. They also offer interpreting equipment rental/sales, video narration, transcription, proofreading/editing and ESL training..

ALTA Language Services,; 3355 Lennox Rd. NE, Ste. 510, Atlanta, GA 30326;
Phone: (404)240_1810 Fax: (404)239_9516; E_mail: robert@altalang.com
Contact: Robert Jones
Company founded 1982. Member: ATA, AAIT. Uses freelance translators and interpreters. Accepts unsolicited resumes. Resumes are filed by language. Requires outstanding translation skills. Main languages (both ways) are Spanish, German, French, Japanese, Portuguese, Dutch, Italian, Chinese, Korean, and Russian. Main subjects are legal, medical, and general. Does multilingual typesetting in addition to translation and interpretation. Engages around 300 translators.

ALTCO Translations, 1426 Ridgeview Rd., Columbus, OH 43221, USA
Phone: (614)486_2014 Fax: (614)559_6682
e_mail: trudypeters@sbcglobal.net
Contact: Trudy E. Peters, Owner.
Founded 1982. Member: ATA. Uses freelancers for translation and interpretation. Only experienced translators' applications accepted. Main languages: all European languages. Main subjects are patents, business, IT, manuals, brochures, advertising, medical/pharmaceutical.

Always Ready Translation Services, 11026 Ventura Blvd., Room 10, Studio City, CA 91604 Phone: (800)240-6601
Fax: (818)755-8959
e-mail: language@worldnet.att.net
Contact: Dan Prescott, President
Founded 1985. Accepts resumes. Translates Spanish, Asian languages, Russian, Armenian, and other European languages.

Ambassador Translating, Inc., 442 Route 202-206 N., #PMB200, Bedminste-r, NJ 07921 Phone: (908)429-0200
Fax: (908)429-0333
e-mail: rosa@am-translating.com Website: www.am-translating.com
Contact: Rosa A. Stevenson Member: ATA

America Translating Services, P.O. Box 800272, Santa Clarita, CA 91380 Phone: (800)535-0555 Fax: (800)316-2230
e-mail: rstevenson@aol.com Website: am-translating.com
Contact: Rosa A. Steventon, Coordinator
Founded 1984. Member: ATA, CCIA. Translates all languages, both from and into English. Emphasis on Legal, Technical and Medical translation. Prospective translators should provide sample translations, reference list. Resumes are listed in a database. Specializes in voiceover work. Maintains a pool of over 1,000 translators.

American Bureau of Professional Translators (ABPT), 8989 Westheimer Rd., Ste. 112, Houston, TX 77063; Phone: (713)789_2500 Fax: (713)789_892-0; E_mail: abpt@abpt.com Web site: www.abpt.com
Contact: Ms. Lu Waterhouse Member: ATA

American Education Research Corp., Box 996, West Covina, CA 91793
Phone: (626)339-4404 Fax: (626)339-9081
e-mail: aerc@cyberg8t.com Website: www.credencialsevaluation.com
Contact: John A. Sheety Member: ATA

American Translation Partners, Inc., 200 Lincoln Street, Boston, MA 02111 Phone 617.350.9988; fax 617.350.6988
E_mail: info@americantranslationpartners.c-om
Web site: www.americantranslationpartners.com
Contact: Scott Crystal
Founded 1998. Member: ATA, NAJIT, NETA, MMIA, LPDAM, FIT, TTI, LISA, AIIC. Accepts resumes. Requires language proficiency and experience as well as computer proficiency. Translates Spanish, Portuguese, French, German, Italian, Japanese, Chinese, Korean, Vietnamese, Haitian Creole, Afrikaans, and all Indian dialects. Works in government, legal, medical, insurance,

technical, financial, software subjects in more than 200 language pairs. Has 1500 translators in New England and 3500 worldwide.

ANDALEX International, Inc., 421 SW 6th Ave. Suite 1150, Portland, OR 97204 Phone: (503)24-.9756 Fax: (503)246-6002
e-mail: info@telelanguage.com www.telelanguage.com
Contact: Andrei Lupenko Member: ATA

i**Antiquariat Literary Services, Inc.**, 12042 Trumbull Way, Reston, VA 20190 Phone:(703)707-8531 Fax:(703)707-8532
e-mail: alstranslation.com Website: www.alstranslation.net
Contact: Rebecca Marie Hindrichs Member: ATA

APS International Ltd; 7800 Glenroy Rd., Minneapolis, MN 55439; Phone: (952)831-7776 Fax: (952)-831-8150
E-mail: trans@civilactiongroup.c-om
Contact: Ann Mickow Member: ATA

Arab American Translators, 5827 Columbia Pike, #415, Falls Church, VA 22041
Phone: (703)820-5612 Fax: (703)820-1318
e-mail: info@translatearabic.com Website: www.translatearabic.com
Contact: Radwan Hakim Member: ATA

ArchiText Inc., 23 Main St., Andover, MA 01810
Phone: (978)409-6112 Fax: (978)409-6096
e-mail: hansf@architext-usa.com Web site: www.architext-usa.com
Contact: Hans Fenstermacher Member: ATA

Argo Translation, Inc., 3747 S. Howell Ave., Milwaukee WI 53207 Phone: (414)615-6000 Fax: (414)615-6005 e-mail: sales@argotrans.com Website: www.argotrans.com
Contact: Jacqueline LaCarelli
Founded 1995. Member: ATA. Uses freelance translators and interpreters. Accepts unsolicited resumes. Works in all languages and subjects, viz., medical, technical, legal, commercial. Always on the lookout for professional translators. Has 575 translators.

Around The World, Inc., 12281 Butterfield Drive NW, Pickerington, OH 43147
Phone: (614)575-2424 Fax: (614)575-2423
e-mail:atw@nls.com Website: www.atwtranslation.com
Contact: Gary West Member: ATA
Founded 1998. Member: ATA, BBB. Translates all languages, mainly technical, advertising, business and legal. Has satellite offices in Kansas City, Columbus, Brussels (Belgium), and Paris (France). Translates and typesets for private and business sectors. Utilizes over 5000 translators.

ASET International Services Corporation, 2009 North 14th St., Ste. 214, Arlington, VA 22201 Phone:(703) 516-9266 Fax:(703)516-9269
e-mail: Khendzel@asetquality.com Website: asetquality.com
Contact: Kevin S. Hendzel, Director of Language Services
Founded 1987. Member: ATA. Translates in over 100 languages, both from and into English. Emphasis on Equipment Manuals, Pharmaceuticals and Medical Equipment, Software Localization, Nuclear, Chemical, Industrial and Structural Engineering, Law and Legislation, Regulations and Codes, Education and Consumer Goods translation. Prospective translators are required to have 10 years professional translating experience and formal technical training in a specialty area. Resumes are archived in database and assignments given when appropriate match arises. Resumes *must be faxed or submitted by e-mail.* The company also works with engineering drawings and specifications and possesses full AutoCad and full-color printing capabilities. They provide interpreting services and arc an authorized Master Distributor for the Philips interpreting equipment series. Welcomes voice-over talent. Maintains a pool of 1000 translators.

Asian Link Corp., 1108 W Valley Blvd., Ste. 4, Alhambra, CA 91803

Phone: (626)300-9191 Fax: (626)300-8955
e-mail:sophia@asialink.com
Contact: Sophia N. Yang   Member: ATA

Asian Translation Service; 7393 S. Locust St., Midvale, UT 84047 Phone: (801)565-8281 Fax: (801)365-6560
E-mail:ats@asiantranslation.c-om Web site: www.asiantranslation.com
Contact: Steve Stevens
Founded 1992. Member: ATA. Translates English into Hmong, Vietnamese, Cambodian, Thai, Korean, Japanese, Chinese, Tagalog, Malay, and Indonesian, mainly health_related, business, personal documents, and product catalogues. Provides full range of translation and interpretation. Uses about 30 translators.

Asian Translations, Inc., 3702 E. Lake St., Ste. 102, Minneapolis, MN 55406
Phone: (612)712-5432 Fax: (612)721-5778
e-mail: atrans@asiantranslations.com
Website: www.asiantranslations.com
Contact: Sokunthea (Sue) Hendry   Member: ATA

ASIST Translation Services, 4663 Executive Drive, Ste. 11, Columbus, OH 43220 Phone: (614)451-6744 Fax: (614)451-1349
e-mail: asist@asisttranslations.com
Contact: Elena Tsinman, President
Founded 1983. Member: ATA. Uses freelance translators and interpreters. Accepts unsolicited resumes. All resumes are entered in a database. If qualified, translator is contacted immediately. Applicant must be experienced, native speaker, and have a university degree. Main languages are English into French, Spanish, Chinese, Japanese, Russian, Arabic, German, Somali, Portuguese and Italian. Additional languages are Dutch, Swedish and Danish. Does 2-3 million words a year. Does also typesetting, Website translation, software translation, localization, audio-visual productions, and desktop publishing. Maintains pool of 2000 translators.

Atlantic International, Translators, Inc., 4956 Vermack Rd., Atlanta, GA 30338
Phone: (770)350-9050 Fax: (770)350-9051
e-mail: atlanticit@aol.com   Website: www.atlanticitinc.com
Contact: Rogelio Cipriano   Member: ATA

Atlas Translation Services, 336 North Central Ave., Ste. 6, Glendale, CA 91203 Phone: (818)242-2400 Fax: (818)242-2475
e-mail: atlas @atlaspvs.com
Contact: Sorina Kalili
Founded 1993. Uses freelancers for translation. Accepts unsolicited resumes. Main languages (both ways) are Farsi, German, Spanish, French, Arabic, Chin-ese, and Japanese. Main subjects are legal, business, and scripts. Also does legal interpretation.

Audio To Go, Inc., 42 West 89th St., New York, NY 10024
Phone: (212)721-1183 Fax: (212)721-1273
e-mail: info@a2g.com   Website: www.a2g.com
Contact: Gayle Goldfarb
Founded 1991. Member: NYCT. Accepts resumes. Requires certification, experience, native speakers in target language. Translates all languages. Translates audio and video programs with narrators in their native language.

Avant Page, 1138 Villaverde Lane, Davis, CA 95616
Phone: (530)750-2040 Fax: (530)750-2024
E-mail: luis@avantpage.com Web site: www.avantpage.com
Contact: Luis Miguel, President
Founded 1996. Member: ATA. Accepts resumes. Requires 3 years exp., certification/accreditation where available. Translates English into Spanish, Japanese, Chinese (simplified and traditional), Korean, French, German, Vietnamese, Arabic, Hebrew, Italian, and Portuguese. Also Farsi, Thai, Tagalog, and MORE!. Main areas are business, hi tech, biotechnology, government documents, and advertisements. Does multilingual desktop publishing. Does over one million words per year. Utilizes over 200 translators.

Avantext, 2991 Shattuck Avenue, Ste. 200, Berkeley, CA 94705
Phone: (510) 644-3490 Fax: (510) 644-3492
E-mail: info@avantext-usa.com Web site: www.avantext-usa.com

-B-

BBC Multilingual Translations, 1231 Somerset Drive, McLean, VA 22101 Phone: (703)448-0893 Fax: (703)448-8077
e-mail: mansourf@aol.com

Babel Tower, Inc., P.O. Box 491023, Ft. Lauderdale, FL 33349
Phone: (954)731-9180 Fax: (954)731-1366
e-mail: babeltower@aol.com Website: www.babeltowerinc.com
Contact: Sandra Fernandez
Founded 1996. Member: ATA. Translates all languages.

Barinas Translation Consultants, Inc., 1100 NW Loop 410, Ste. 700, San Antonio, TX 78213 Phone: (210)545-0019 Fax: (210)545-4731 160424 E-mail: info@barinas.com Website: www.barinas.com
Contact: Sonia Barinas, President
Company founded 1980. Member: MPI, AAHA, TSHE, SATC, San Antonio Chamber of Commerce, GSHMA, ISMP, THMA. Uses freelance translators and interpreters. Accepts unsolicited resumes. Resumes are checked for quality and experience, then filed by language and areas of expertise. Please submit sample with resume, and indicate degree in translation and/or interpretation. Degree in law, medicine etc. is a plus. Main language pairs are Spanish/English, French/ English, French/Spanish, Portuguese/ English, German/English, Chinese/English, Japanese/English. Main subjects are legal, medical, technical, telecom. Specializes in simultaneous interpretation for meetings and conventions. Does about 350 projects a year. Uses hundreds of freelancers.

Berkeley Scientific Translation Service, Inc, P.O.Box 150, Berkeley CA, 94701; Phone: (510)548_4665 Fax: (510)548_4666; E_mail: marlo@berksci.c-om Web site: www.berksci.com
Contact: Dr. Marlo R. Martin
Founded 1974. Member: ATA. Translates Japanese, Korean, Chinese, and other major European and Asian languages, both from and into English. Emphasis on Mechanical, Automotive, Chemical and Chemical Engineering, Computers and Software, Electronics, Biotechnology, Pharmaceutical-s, Physics and Patent translations. Prospective translators should be able to produce autho_ritative translations into their native language within a specialized area. Resumes are screened for educational background and subject expertise as related to translation experience. The company was founded by a physicist and engineer and seeks translators with similar qualifications. Maintains a pool of 70-100 translators.

Betmar Languages, 6260 Highway 65 NE, Ste. 308, Minneapolis, MN 55432 Phone: (612)572-9711 Fax: (612)571-3467
e-mail: best@betmar.com Contact: Elizabeth A. Loo Member: ATA

BioMedical Translators, 3477 Kenneth Drive, Palo Alto, CA 94303 Phone:(650)494-1317 Fax:(650)494-1394
e-mail: biomed@biomedical.com Contact: Monique Vazire
Founded 1992. Member: ATA, NCTA. Main languages translated are French, German, Italian, Spanish, Dutch, Swedish, Portuguese, Danish, Japanese, and Chinese. Emphasis is on Medical and Biological translation, including equipment, studies and software. Prospective translators should possess at least one year experience, have knowledge of the medical field, and have access to medical dictionaries. Their equipment should include a PC and modem or e-mail; software should include Word and WordPerfect. Resumes are reviewed by the recruiting department and responded to with a test translation, which is then evaluated. The company specializes exclusively in the medical field and its peripherals. Their services also include desktop publishing. Maintains a pool of 500+ translators.

Bloomberg L.P., 100 Business Park Drive, P.O. Box 888, Princeton, NJ 08542 Phone:(609)279-4955 Contact: Elsa Shilling

Bowne Translation Services, 345 Hudson Street, New York, NY 10014
Tel. (212) 229-7370 Fax. (212) 229-7395
Email: lisa.dimeglio@bowne.com

BRADSON Corporation, 1735 Jefferson Davis Highway, Ste. 705, Arlington, VA 22202 Phone: (703)413-3050 Fax: (703)413-3056
Contact: John S. Horner

BRIDGE-LINGUATEC LANGUAGE SERVICES, 915 South Colorado Blvd., Denver, CO 80246 Phone: (303)777-7783 Fax:(303)777-7246
e-mail: translators@bridgeschool.com
Website: www.bridge-linguatec.com Contact: Michelle Duvall
Founded 1981. Translates mainly Spanish, French, German, Arabic, Dutch, Russian and Portuguese. Emphasis on legal and medical. Prospective translators should, by preference, be native speakers. Resumes are filed by language and translators contacted by need.. The company also provides ESL training for executives, foreign language instruction, and interpretation services. Maintains a pool of 150-200 translators.

Brigham and Women's Hospital, 75 Francis St., Boston, MA 02115
Phone: (617)732-6640 Fax: (617)975-0813 e-mail: igarcia@partners.o-rg
Contact: Ileana Jimenez Garcia Member: ATA

Bromberg & Associates, 28475 Ranchwood Drive, Southfield, MI 48076
Phone: (248)827-7164 Fax: (248)827-3621
E-mail: jinny@linguacity.com Web site: www.linguacity.com
Contact: Jinny Bromberg Member: ALC
Bruce International, Inc., 10550 SW Allen Blvd., Ste. 123, Beaverton, OR 97005
Phone: (503)643-8448 Fax: (503)643-7174
E-mail: info@bruceinternational.com
Web site: www.bruceinternational.com
Contact: Oscar Andrino Member: ATA

Burg Translation Bureau, Inc., 29 South LaSalle, Ste. 936, Chicago, IL 60603 Phone: (312)263-3379 Fax:(312)263-4325
E-mail: burg@interaccess.com Web site: www.burgtranslations.com
Contact: Lodovico Passalacqua, President
Founded 1936. Member: ATA. Emphasis on technical translation. Translators carefully "screened" _ Strong Project Management - QC procedure en-forced.Translati-on and Typesetting.

-C-

CC Scientific, Ltd., 73920 Mountain View Ave, Palm Desert, CA 92260
Phone: (760)341-7544 Fax: (760)341-7514 e-mail: CCSLtd@aol.com
Contact: Dr. Jaime R. Carlo-Casellas Member: ATA

C.P. Language Institute, 225 West 57th St., Ste. 404, New York, NY 10019 Phone: (212)246-2054 Fax: (212)247-2258
e-mail: info@cpli.com Website: www.cpli.com
Contact: Yuki Saito
Founded 1980. Member: ATA. Translates Chinese, Japanese, Korean, Spanish, Arabic, Russian, French, German, Hebrew, Hindi, Urdu, Vietnamese, Italian, Portuguese, Thai, Khmer, Tagalog and Indonesian, both from and into English. Emphasis on Advertising, Technical, Patent, Financial, Medical, Legal, Personal Document and Textbook translation. Prospective translators should show pro-ficiency in English and the target language and should be experienced. Resumes are responded to with a call-in or an e-mailed test translation in order to assess the candidate's ability. The company has two departments - programs and services. Programs is a language school providing small group and private language instruction. Services comprise translation, typesetting, formatting, interpretation, voiceovers and subtitling. Maintains a pool of 300 translators.

Carolina Polyglot, Inc., PO Box 36334, Charlotte, NC 28236

Phone:(704)366-5781 Fax:(704)364-2998
e-mail: wdepaula@carolinapolyglot.com
Website: www.carolinapolyglot.com Contact: Dr. William DePaula
Company founded 1971. Member: ATA, CATI. Accepts unsolicited resumes. Resumes are filed by language pair. E-mail and fax filed electronically. Require-ments for applicants include academic degree, previous experience, professional affiliation/ accreditation, references. Main languages (both ways) are French, Spanish, Italian, Portuguese, Romanian, German, Dutch, Arabic, Chinese, Viet-namese, and Japanese. Also translates Turkish, Farsi, Africaans, Hindi, Danish, Norwegian, Swedish, and Finnish. Main subjects are immigration, education, law, business, medicine, insurance, theology, ecology, literature, and computers. Uses over 30 translators.

Caterpillar, Inc., Dealer Capability Dept., Corp Translation, 501 SW Jefferson Ave., Peoria, IL 61630
Phone: (309)494-5216 e-mail: opherk_jorg@CAT.com Member: ATA
Contact: Dr. Jorg Opherk Member: ATA

CDNOW, Inc., 1005 Virginia Drive, Fort Washington, PA 19034
Phone: (215)619-9696 Fax: (215)619-9559
e-mail: emcpartland@cdnow.com Website: www.cdnow.com
Contact: Erica A. McPartland Member: ATA

Certified Interpreters & Translators (CIT), PO Box 390006, San Diego, CA 92149 Phone: (619)475-8586 Fax: (619)472-2157
Company founded 1983. Uses freelance translators and interpreters. Accepts unsolicited resumes. Resumes are filed by language and geographical area. Interpreters have to be certified. Main language pair is English-Spanish, followed by English into German and Tagalog, Russian, Arabic, and Japanese into English. Also Portuguese, Polish, and Vietnamese into English. Main subjects are legal, medical, technical, literary, commercial, as well as manuals. Does conference interpreting. Also does translator and interpreter training.

Certified Languages International, 4700 SW Macadam Ave., #200, Portland, OR 97201 Phone: (503)525-9601
Fax: (503)525-9607
E-mail: Kristin@certifedlanguages.com
Web site: www.certifiedlanguages.c-om
Contact: Kristin Quinlan Member: ATA, ALC

Certified Translation Services, One Harbison Way, Ste. 105, Columbia, SC 29212 Phone: (800)730-9970 Fax: (803)781-5052
e-mail and website: certifiedtranslationservices.com
Contact: Ed Crosby Member: ATA

CETRA, Inc; 4312 Oak Avenue, Melrose Park, PA 19027; Phone: (215)635-7090 Fax: (215)635-9239; Web Page: www.cetra.com E-mail: info@cetra.com;
Contact: Dr. Jiri Stejskal, President; Member: ATA
Founded in 1994. The main languages that they translate are English into French, German, Spanish, Japanese and Chinese. They also translate Central/East European languages. The subjects that they translate are business, legal and surveys. CETRA does use freelance and they do accept resumes. The resumes are reviewed and filed together with information provided by the vendor via online form. You may fill out a form at http://www.cetra.com/trans_appl.htm-l.

Cherry Creek Language Centre, Inc., 1071 Jersey St., Denver, CO 80220
Phone: (303)756-2520 Fax: (303)782-5686
e-mail: ignacio@translationlinks.com Web site: www.translationlinks.com
Contact: Ignacio Jimenez (DBA.Translationlinks tradename.)

Choice Translating & Interpreting, 8701 Mallard Creek Rd., Charlotte, NC 28262 Phone: (704)717-0043 Fax: (704)717-0046
e-mail: hr@choicetranslating.com Website: www.choicetranslating.com
Contact: Sophie David-Gunn Member: ATA

CinciLingua, Inc., 322 East Fourth St., Cincinnati, OH 45202 Phone: (513)721-8782 Fax:(513)721-8819
email: translat@cincilingua.com Website: www.cincilingua.com
Contact: Michael Sum
Founded 1972. Member: ATA, ASME. Translates in Spanish, French, German, Chinese, Japanese, etc., both from and into English. Prospective translators should have college degree, professional experience in a technical specialty, and references. Should have a computer. Resumes are reviewed by the Project Manager and entered in a database. Qualified candidates may be asked to translate a short sample for additional information.

CITI Translation Center, Inc., 1025 Kane Concourse, Ste. 208, Bay Harbor Islands, FL 33154 Phone: (305)868-1746 Fax: (305)868-1748
e-mail: info@cititran.com Website: www.cititran.com
Contact: Fernando Velez-Pardo
Founded 1987. Member: ATA, FLATA. Translates German, Russian, Japanese, Chinese, French, Italian and Spanish, both from and into English. Emphasis on Technical, Communication Manuals, Medical, Business and Personal Documents translation. Prospective translators should be experienced, preferably with accre-ditation and/or certification. Resumes are reviewed and the applicant contacted if there is a specific need for that language or specialty area. The company supports markets in the Americas, Europe and the Pacific Rim. Maintains a pool of 60 translators.

CME Corporation, 2945 Three Leaves Drive, Mount Pleasant, MI 48858 Phone: (517)773-0377 Fax: (517)773-1105
Contact: Fumiko Lobert Member: ATA

Columbia Language Services, 11818 SE Mill Plain Blvd., #307,
Vancouver, WA 98684 Phone: (360)896-3881 Fax: (360)896-4074
e-mail: mail@columbia-language.com
Website: www.columbia-language.com
Contact: Sveltana Linchuk Member: ATA

CommGap International Language Services
PO Box 71248, Salt Lake City, Utah 84171
Phone: (801) 944-4049 Fax: (801) 944-4046
e-mail: info@commgap.com Web site: www.commgap.com
Contact: Lelani P. Craig Member: ATA

Community Interpreter Services, Catholic Charities/Greater Boston, 75 Kneeland St., 8th Floor, Boston, MA, 02111
Phone: (617)350-4811 Fax: (617)629-5768
e-mail: inna_persitsgimelberg@ccab.com Web Page: www.cccis.org
Contact: Inna P. Gimelberg, Program Manager
Founded in 1986. Member: ATA, MMIA, NETA. Accepts resumes. Translates Haitian Creole, Spanish, Cape Verdean Creole, Vietnamese, Russian, Portuguese, Cantonese-Chinese, Vietnamese, Polish, Farsi, Bosnian and Albanian. Also Somali, Swahili, Italian, French, Arabic, Khmer and Laotian. They translate a total of 50 languages. Emphasis on Personal Document translation. Prospective translators and interpreters should live in Massachusetts or a bordering state, and possess at least a Bachelor's degree, past translating experience, and fluency in at least two languages. Qualified resumes are responded to with an interview appointment. They are a nonprofit service operated by Catholic Charities. Maintains a pool of 200 translators and interpreters.

ComNet International, 31255 Cedar Valley, Ste. 212, Westlake Village, CA 91362 Phone:(818)991-1277 Fax:(818)991-1699
Website: comnetint.c-om Contact: Dr. Elias Agel
Founded 1989. Member: ATA. Translates all languages, in all subjects. Pro-spective translators should have extensive experience and should provide references. Resumes are reviewed and entered into a database. The company also provides desktop publishing, art production and printing services. It also does voice-overs and dubbing. Specialties include Middle East consulting services, electronic update filing and interagency network cooperation. Maintains a pool of 600 translators.

Comprehensive Language Center, 4200 Wilson Blvd., Ste. 950, Arlington VA 22203 Phone: (703)247-0700 Fax: (703)247-4295

e-mail: kjdresen@comlang.com Website: www.comlang.com
Contact: Kelly Jones Dresen
Company founded 1980 as CACI. Member: ATA. uses freelance translators and interpreters. Accepted unsolicited resumes, but not unsolicited phone calls. Qualified applicants are sent a database form which is processed and are called upon when needed in their area of expertise. Requirements include college degree, 2 years professional experience as linguist, or appropriate certification/ specialized degree. Over 100 languages. Main areas are technical, legal, promotional, business. Company also does training, transcription, software and web page localization, and video narration.

Continental Communications Agency, 5105 E. Los Angeles Ave. #145
Simi Valley, CA 93063 Phone:(805)527-4446 Fax:(805)527-4460 e-mail:
cca1@adelphia.net or cca_inc@ix.netcom.com
Contact: Peter Charbonne-au Member: ATA

Continental Interpreting Services, Inc., 3111 N. Tustin Ave., Ste. 235,
Orange, CA 92865 Phone:(800)201-7121 Fax:(714)283-9045
e_mail: veronica@wespeakyourlanguage.com
Web site: www.wespeakyourlanguage.com
Contact: Veronica Amador Member: ATA

Copper Translation Service, 18 Joseph Lane, Gansevoort, NY 12831
Phone: (518)691-0155 Fax: (518)691-0157
e-mail: angelos@coppertranslation.com
Web site: www.coppertranslation.com
Contact: Angelos G. Tzelepis
Company founded 1971. Member: ATA. Uses freelance translators and interpreters. Accepts unsolicited resumes. Applicant must be native speaker of target language. Main areas are technical legal, medical, and business. Over 700 translators in pool.

Corporate Translations, 1300 Aviation Blvd., Redondo Beach, CA 90278
Phone: (860)450-0405 Fax: (860)450-0409
e-mail:ctoomey@corptrans.inc.com Website: www.orptrans.inc.com
Contact: Carolyn Toomey
Founded 1995. Member: ATA

Cosmopolitan Business Communication, 8025 Marshall Circle, Arvada, CO 80003 Phone: (303)422-6702 Contact: Maya León Meis

Cosmopolitan Translation Bureau, 53 West Jackson Blvd., Ste. 1260, Chicago, IL 60604 Phone: (312) 726-2610 Fax: (312)427-8591
Contact: Emanuel H. Steen Member: ATA

Coto Interpreting, Translating & Graphics, 447 Burchett St., 2nd Floor, Glendale, CA 91203 Phone: (818)551-4545 Fax: (818)551-1123
Contact: Melanie Coto-Trevor Member: ATA

Course Crafters, 44 Merrimac St., Newburyport, MA 01950
Phone: (978)465-2040 Fax: (978)465-5027
E-mail: LBRagan@coursecrafters.com Web site: www.coursecrafters.com
Contact: Lise B. Ragan Member: ATA

Creo International, 520 Marquette Ave., Ste. 700, Minneapolis, MN 55402
Phone: (612)342-9800 Fax: (612)342-9745
e-mail:cthudson@creo-works.com Website: www.creo-works.com
Contact: Christa Tiefenbacher Member: ATA

Counterpoint Language Consultants, Inc., P.O. Box 6184: Bridgewater, NJ 08807 Phone: (908)231-0991 Fax: (908)231-266
E-mail: ctpt@ctpt.net Web Page: www.ctpt.net
Contact: Erik Greb
Founded 1979. Member: ATA. Company uses freelance translators and interpreters. Accepts unsolicited resumes. Main languages (both ways) are French, German, Spanish, Chinese, Japanese, Portuguese, Italian, Russian, Greek and Hindi. Also Thai, Gujarati and Slovak. Main Areas are pharmaceuticals, telecom and legal. Provides worldwide language training.

Cross Cultural Communications Systems, Inc: 10 Converse Place, P.O. Box 860, Winchester, MA 01890 Phone: 781-729-3736 ext.110
Fax: 781-729-1217 E-mail: vphillips@cccsorg.com
Contact: Vonessa Phillips; Member: ATA

Crossword Translation & Interpreting, 16350 Park Ten Place, Ste. 201, Houston, TX 77084 Phone: (281) 578-1300 Fax: (281) 578-1334
e-mail:language@crosswordtranslation.com
Contact: Office Staff
Founded 1997. Member: ATA. Accepts resumes. Translates and interprets most languages in all fields. Uses many translators and interpreters.

Cybertec USA, Inc., 153 W. Westfield Ave., Roselle Park, NJ 07204-1816
Phone: (908)245-3305 Fax: (908)245-5434
E-mail: mail@cybertecusa.com Web site: www.cybertecusa.com
Contact: Joseph Nunes
Technical Translation Bureau founded in 1990. Member: ATA, NYCT. Uses in-house and freelance professional translators. Accepts unsolicited resumes. Resumes are filed by language and specialty. Translators are required to be experienced, professional translators, preferably with language or other diploma and accreditation/certification. Translates mainly Portuguese (Euro, African and Brazilian) and Western European languages. Specialized in technical and commercial subjects, including many industries, petroleum, biotech. Does over 4 million words per year. Over 100 translators in company's pool. Offices in the USA and Portugal.

CyraCom International Inc., 7330 N Oracle Rd., Tucson, AZ 85704
Phone: (520)745-9447 Fax: (520)745-9022 Website: www.cyracom.com
Contact: Paul Burns
Company founded 1997. Member: ATA. Provides telephonic inter-pretation services on demand 24 hours a day, 7 days a week in about 150 languages. Has a secure system that provides rapid access to interpreters.

-D-

DFP Translating Service, 42 Frederick Place, Bergenfield, NJ 07621
Phone: (201)343-2258 Fax: (201)343-4155
e-mail: info@ translationservices.net
Website: www.translationservices.net Contact: Doris Pacheco

DTS Language Services, Inc., 308 W. Rosemary St., Ste. 210, Chapel Hill, NC 27516 Phone: (800)524-0722 Fax: (919)942-0686
e-mail: sales@dtstrans.com Website: www.dtstrans.com
Contact: Lucia Apollo Shaw
Founded 1972. Member: ATA, DIA, STC. Accepts resumes only electronically. Requires 5 years min. professional technical translation experience with related degree. Translates mainly Portuguese, French, Spanish, German, Italian, Czech, Vietnamese, Japanese, and Korean. Also Dutch, Finnish, Norwegian, Swedish. Mainly technical, medical, marketing, patents.

Detroit Translation Bureau, 30800 Telegraph Rd., Ste.1930, Bingham Farms, MI 48025 Phone: (248)593-6710
Fax: (248)593-6720

e-mail: info@dtbonline.com   Website: www.dtbonline.com
Founded 1946. Member: ATA, LISA, SAW. Accepts resumes. Translates mainly Spanish (Mexico, Spain), Canadian French, German, Japanese, Italian, Norwegian, Dutch, Finnish and Korean. Mainly automotive – technical, training and marketing. Uses hundreds of translators.

Digital Publishing, Inc., 1805 Hicks Rd., Rolling Meadows, IL 60008
Phone: (847)776-0200   Fax: (847)776-0249
e-mail: translation@digitalpub.com   Website: www.digital pub.com
Contact: Debbie Shulman

Digitek International, Inc./SAKHR, 7631 Leesburg Pike, Ste. B, Fall Church, VA 22043   Phone: (703)883-0134
Fax: (703)883-0137
Contact: Youb Benyoucef
Diplomatic Language Services, 1901 N. 19th St., Ste. 600, Arlington, VA 22209 Phone:(703)243-4856 Fax:(703)358-9189
e-mail: resume@dls-inc.com   Website: www.dls-inc.com
Founded 1985. Member: ATA, NCATA, NAJIT. Accepts resumes. Requires 5 years professional experience. Translates all languages. Provides all translation and interpretation services as well as language training.

DocuTrans, Inc., 4712 S. 2675 W, Roy, UT 84067
Phone: (801)916-3924 Fax: (801)776-4027
e-mail: translate@docutrans.com   Web site: www.docutrans.com
Contact: Gabariele H. Johnson   Member: ATA

Dragon Crest Associates, Inc., 11026 Ventura Blvd., Ste. 10, Studio City, CA 91604   Phone: (800)240-6601 Fax: (818)755-8959
e-mail: language@worldnet.att.com
Contact: Dan Prescott   Member: ATA

-E-

800-Translate, 865 United Nations Plaza, New York, NY 10017
Phone: (800)872-6752   Fax: (888)872-6752
Website: www.800-translate.com   Contact: Ken Clark

echo international, Three Gateway Center, Floor 14 W, Pittsburgh, PA 15222   Phone: (412) 261-1101   Fax: (412) 261-1159
e-mail: dutka@echotrans.com   Contact: Leslie A. Dutka

Echo Translation, P.O. Box 15301, Minneapolis, MN 55415
Phone: (612) 822-3473   Fax: (612)339-2122 e-mail: echobea@aol.com
Contact: Bea Cabrera

Eiber Translations 65-11 174th St., Fresh Meadows, NY 11365
Phone: (718) 463-2900   Fax: (718) 359-4073 e-mail: ibertrans@aol.com
Contact: Edna H. Eiber

El Andar Translations, P.O. Box 7745, Santa Cruz, CA 95061
Phone: (831) 457_8353 Fax: (831) 457_8354
e-mail: translations@elandar.com   Web site: www.elandar.com
Contact: Marga Vaquer

eLocale, Inc., 815 Park Blvd., Ste. 255, Boise, ID 83712
Phone: (208)336-5530   Fax: (208)336-6445info@

elocale.com Web site: www.elocale.com
Contact: Cary Clark

Eriksen Translations Inc., 32 Court St., 20th Fl., Brooklyn, NY 11201
Phone: (718) 802-9010 Fax: (718) 802-0041
e_mail: info@eriksenine.com Web site: www.eriksenine.com
Contact: Leah Ruggiero
Founded 1986. Member: ATA, NYCT, STC. Translates into and from all major languages. Emphasis on Legal, Financial, Advertising, Education, Pharmaceutical and Health Care. Translators are encouraged to register on Web site. The company also provides typesetting services and Web site globalization. Maintains a pool of 3000 translators.

eTranslations, 1930 Village Center Circle #3-930, Las Vegas, Nevada 89134
Phone: 1-888-262-3468 (USA only) / (303)388-4488 Fax: (303)388-2255
E-mail: info@etranslations.com Web site: www.etranslations.com
Contact: Desiree Blum

Eurasia International, 16530 Ventura Blvd., Ste. 206, Encino, CA 91436 Phone: (818)907-9718 Fax; (818)907-9763
e-mail: VMordukhay@worldnet.att.net Website: www.eurasia-usa.com
Contact: Violetta Mordukhay
Founded 1993. Member: ATA. Accepts resumes. Translates mainly Spanish, German, Hebrew, Russian, French, Arabic, Farsi and Indonesian. Also Romanina, Slovak, Malay, Japanese, Polish, Czech and Chinese. Mostly immigration, legal, and book publishing. Pool of over 200 translators.

Eureka-Foreign College Evaluators & Translators, 6601 W. Irving Park Rd., Ste. 204, Chicago, IL 60634
Phone:(773)545-1700 Fax:(773)545-1716
e-mail: eu-reka@interaccess.com Website: www.polishtranslations.c-om
Contact: Maria Drewicz, Executive Director
Founded 1992. Member: ATA, The Translators and Interpreters Guild. Translates Polish, both from and into English, and German and Spanish into English. Emphasis on University Admission Documents, Immigration & Naturalization, Business, Technical, Medical and Legal translation. Prospective translators should have IBM-compatible PC, Word Perfect 5.1 DOS or Windows, minimum Bachelor's Degree. Resumes are reviewed, and qualified persons contacted. The company specializes in credential evaluation and Polish-American consulting services for business cooperation, cultural and marketing advisement, and is currently looking for Polish translators and interpreters. Maintains a pool of 10 translators.

EuroNet Language Services, 295 Madison Ave., 45th Fl., New York, NY 10017 Phone:(212)271-0401 Fax:(212)271-0404
e-mail: euronet@mindspring.com
Contact: Anouk, HRM
Founded 1989. Member: NYCT, ATA. Uses freelance translators and interpreters. Accepts unsolicited resumes. Translates Spanish, French, Portuguese, Italian, Dutch, German, Swedish, and other European languages, both from and into English. Translates all subjects. Prospective translators must be native speakers with bilingual capacity in English and possess a university degree. Resumes are reviewed and added to a database. Maintains a pool of 300 translators.

Eurotext Translations Rosetta Stone, 34 Franklin St., Ste. 200A, Nashua, NH 03063 Phone: (603)883-9388 Fax: (603)595-8673
e-mail: translate@rosettastoneinc.net
Web site: www.etrosettastoneinc.c-om
Contact: John Furey, CVO
Founded 1972. Member: GALA, STC, NHITA. Accepts resumes. Translates all major languages including French, Italian, German, Spanish, Russian, Chinese, Japanese, Arabic, Portuguese and Greek. Mostly industrial, medical, legal, scientific, marketing,

computers, software and localization. Also provides the following services: interpretering, voice_overs, localization, language instruction and cross-cultural training. Aquired by Eurotext Translations, Dublin, Ireland in October 2002.

Excel Translations, Inc., 660 Market St., Ste. 206, San Francisco, CA 94104 Phone: (415)434-4224 Fax: (415)434-4221
e-mail: info@xltrans. com Web site: www.xltrans.com
Contact: Hervé Rodriguez
Founded 1995. Offices in California, Pennsylvania and Spain. Member: ATA, NCTA, STC, HTML Guild. ISO 9001:2000 certified. Translates into and from all major European and Asian languages. Emphasis on Software Localization and Life Sciences. Prospective translators should have three years' translation experience and a specialty field. They should be native speakers and have access to the Internet. Most translations done using TM tools. Applicants will need to take a test. Resumes are screened and entered into a database if test is successful. Applicants are notified by postcard. The company works with both freelancers and in_house translators and provides full DTP services. Maintains a pool of 1,000 translators.

Executive Linguist Agency, Inc., 500 S. Sepulveda Blvd., Ste. 300, Manhattan Beach, CA 90266 Phone: (310)376-1409 Fax (310)376-9285 e-mail: mail@executivelinguist.com Website: www.executivelinguist.c-om Contact: Ronald R. Randolph

Expert Translators, 10621 N. Kendall Drive, Ste. 111, Miami, FL 33176. Phone: (305)279-8353 Fax: (305)273-4407
e-mail: experttranslators.com Website: www.experttranslators.com
Contact: Emily Correal

-F-

FLS, Inc., 3609 A-5 Memorial Parkway SW, Huntsville AL 35801
Phone: (205)881-1120 Fax: (205)880-1112
e-mail: mail@flstranslation.com Website: www.flstranslation.com
Contact: Judith H. Smith
Company founded 1979. Member: ATA. Uses freelance translators and interpreters. Unsolicited resumes accepted. Translates over 60 languages, mainly Japanese, Spanish, French, Chinese, Korean, German and East European languages. Mostly technical and legal. Over 200 translators.

Federal News Service, 620 National Press Building NW, Washington, DC 20045 Phone: (202)347-1400 Fax: (202)626-2630
e-mail: info@fednews.com Website: www.fednews.com
Contact: Marina Makanova
Founded 1987. Accepts resumes. Requires college degree. Translates mainly Spanish, French, Russian, Chinese, and Arabic. Uses 63 translators. The company is a news wire that provides government transcripts, online newsclipping, news monitoring services, and translations in almost any language.

First Translation Services, 3072 Brighton, 1st St., Brooklyn, NY 11235
Phone: (718)372-7723 Fax: (718)265-6653
Contact: Jack Kleyman Member: ATA

Foreign Exchange Translations, Inc., One Richmond Square, Providence, RI 02906 Phone: (401)454-0787 Fax: (401)454-0789
e-mail: adnresh@fxtrans.com Website: www.fxtrans.com
Contact: Andres Heuberger Member: ATA

Foreign Ink, Ltd., 5735 Washburn Ave. S, Minneapolis MN 55410 Phone:
612)920-4884 Fax: (612)924-9075
e-mail: greg@fornink.com Web site: fornink.com

Contact: Gregory Stricherz Member: ATA

-G-

GES Translation Services, 836 Manhattan Ave., Brooklyn, NY 11222
Phone: (718)389-8453 Fax: (718)389-4442 e-mail:ges@flashcom.net
Contact: Richard Rzeznik
Garcia-Shilling International, 1402 Corinth, L.B. 139, Dallas, TX 75215
Phone: (214)428-4428 Fax: (214)428-4458
e-mail: acento@swbell.net Contact: Luis A. Garcia, Sr.

Gene Mayer Associates, 9 Depot St., 2nd Fl., Milford, CT 06460
Phone: (203)882-5990 Fax: (203)882-5995
e-mail: info@design4language.com
Founded 1985. Member: ATA, AIGA, Design Management Institute, National Investor Relations Institute, CT Art Directors Club. Provides graphic design, translation and production of complex printed material used by global corporations. Emphasis on business (employee information, annual reports, marketing brochures, corporate policy manuals). The firms handles translation, design, book layout and production in 30 languages. Prefers to work with agencies or with translators with a qualified track record with large corporate clients.

Geneva Worldwide Inc; 62 West 22nd Street, New York, NY 10010 Phone: (212)255-8400 Fax: (212)255-8409;
E-mail: genevaworldwide@aol.com
Web Page: www.genevaworldwide.c-om
Contact: Craig Buckstein, VP Operations.
Member: GSA, ATA, NAJIT, RID, NYCT, OGS. Geneva Worldwide was founded in 1903. They translate Spanish, Creole, Chinese (T+S), Arabic, Polish, Korean, Bengali, Urdu, Russian, Albanian. In addition they also translate 180 more languages. Their main subjects are business, medical, legal, general, technical, and financial. They specialize in various types of translation. They do accept freelance along with resumes. The resumes are filed by language and coded by experience and SME. Requirements for applicants are professional resumes. Geneva Worldwide provides face to face interpretation and document translation in over 180 different languages. They provide multilingual services to many Federal, State, City and Private agencies. They have several hundred translators in their pool.

The Geo Group, 6 Odana Court, Ste. 205, Madison, WI 53719
Phone: (608)230-1000 Fax: (608)230-1010
e-mail: translation@thegeogroup.com Website: www.thegeogroup.com
Contact: Translation Manager
Founded 1991. Accepts resumes. Requires min. 3 years translation/inter-preting experience. Main languages are French, German, Hmong, Italian, Portuguese, Spanish, Japanese, Chinese and Dutch. Also other European languages. Main subjects are medical, automotive, marketing, computers, dairy equipment, legal and audio/video. Uses a pool of over 500 translators.

Geotext Translations, 259 W. 30th St. 17th Floor, New York, NY 10001
Phone: (202)631-7432 Fax: (202)631-7407
e-mail: translations@geotext.com Web site: www.geotext.com
Contact: Randon Burns

German Language Services, 2658 48th Ave., SW, Seattle, WA 98116
Phone: (206)938-3600 Fax: (206)938-8308
e-mail: GermanLanguage@GermanLanguageServices.com
Web site: www.GermanLanguageServices.com
Contact: Maia Costa
Founded 1979. Member: ATA, NOTIS, GACC. Accepts resumes only in electronic form. Requires formal training in translation or interpretation and 5 years experience. Only translates German.

Global Advanced Translation Services, Inc., 501 S. Fairfax Ave., #212, Los Angeles, CA 90036
Phone: (323)937-1525 Fax: (323)937-6568 e-mail: mteper@aol.com
Contact: Marina Teper Member: ATA
Founded 1979. Member: ATA, NOTIS, GACC. Accepts resumes in electronic form. Requires training in translation or interpretation and 5 years experience. Only translates German.

The Global Institute of Languages and Culture, Inc., 7301 NW 4th st., Ste. 110, Plantation, FL 33317 Phone: (954)327-1662 Fax: (955)327-8116 e-mail: globalin@gate.net Website: www.globalinstitute.com
Contact: Antonieta Mercado
Member: ATA. Translates Spanish, Portuguese, Italian, French, German, etc., both from and into English. Emphasis on General, Computer, Medical and Legal translation. Prospective translators should have a college degree, experience, and should be native speakers of the target language. Resumes are responded to based on the applicant's qualifications and the company's needs. Maintains a pool of 20 translators.

Global Languages & Cultures, Inc., 680 N. Lake Shore Drive, Ste. 919, Chicago, IL 60611 Phone: (312)981-4444
Fax: (312)981-4445
e-mail: yozturk@e-translation.com Website: www.e-translation.com
Contact: Sedef Olcer

Global Language Solutions/The Russian Word, Inc., 19800 MacArthur Blvd., Ste. 520, Irvine, CA 92612
e-mail: info@globallanguages.com Website: www.globallanguages.com
Contact: Inna Kassatkina

Global Translation Systems Inc., 910 Airport Rd., Chapel Hill, NC 27514 Phone:(919)967-2010 Fax:(919)929-1333
e-mail: mike@globaltranslation.com Website: www.globaltranslation.com
Contact: Michael Collins Member: ATA

Global Visions LLC, 900 Commonwealth Place, Suite # 218., Virginia Beach, VA 23464 Phone: (757)495-3719
Fax: (757)495-6398
e-mail: globalvisions48@msn.com
Contact: Kristi B. Emerson Member: ATA

The Global Word, Inc., 63 Grand Ave., River Edge, NJ 07661
Phone: (201)343-0015 Fax: (201)343-4155
e-mail: info@globalword.com Website: www.globalword.com
Contact: Michael Fundaro Member: ATA

Global Works.Com, 129 W. 27th St., New York, NY 10001
Phone; (212)414-1300 Fax: (212)414-8700
e-mail: emantello@globalworks.com Website: www.globalworks.com
Contact: Elisabeth Mantello Member: ATA

Go Global Translations, Inc., 19515 Rosedale Ave., Excelsior, MN 55331 Phone: (952)474-6413 Fax: (952)474-3756
e-mail:TeresaGoGlobal@cs.com Website: www.goglobalranslations.com
Contact: Teresa A. Crespo Member: ATA

The Golf Channel, Inc., 7580 Commerce Center Drive, Orlando, FL 32819 Phone: (407)248-3344 Fax: (407)248-1344
e-mail: jkrisand@tgcinc.com

Contact: James R. Krisanda Member: ATA

-H-

H&K Translations, LLC, 701 Brickell Ave., Suite 3000, Miami, FL 33131 Phone: (305)789-7521 Fax: (305)789-7799
Contact: Maury M. Scharrer Member: ATA

Health Outcomes Group, 3600 West Bayshore Road, Suite 106, Palo Alto, California 94303 Tel: (650) 858-8033 Fax: (650) 858-8034
e-mail: david_himmelberger@healthoutcomesgroup.com
www.healthoutcomesgroup.com

Hernandez & Garcia, LLC, 7360 N. Lincoln Ave., Ste. 100, Lincolnwoo-d, IL 60712 Phone: (847)676-4445 Fax: (847)676-1420
e-mail: mhernandez@hernandezgarcia.com
Contact: Maritza Hernandez Member: ATA

Hightech Passport Limited, 1590 Oakland Rd., Ste. B202, San Jose, CA 95131 Phone: (408)453-6303 Fax: (408)453-9434
e-mail: magdalena@htpassport.com Website: www.htpassport.com
Contact: Magdalena Enea, VP of Market Development
Founded 1992. Member: ATA, NCTA. Translates French, German, Spanish, Italian, Portuguese, Swedish, Dutch, Russian, Japanese, Chinese, and all Scandinavian from English. Emphasis on Computer Software, Documentation (both printed and online), Marketing Materials, Tapes, and Computer Training Literature translations. Complete range of internationalization and localization services, mainly to IT and medical industry. The company also provides software localization and testing, videotape localization, and desktop printing services. Maintains a pool of 150 translators.
Honda R&D North America, Inc., 21001 State Route 739, Raymond,
OH 43067 Phone: (513)645-6164 Fax: (513)645-6341
e-mail: knason@oh.hra.com Contact: Kay T. Nason

-I-

IDEM Translation, 550 S. California Ave., Ste. 310, Palo Alto, CA 94306
Phone: (650)329-0170 Fax: (650)329-0266
e-mail: idem@compuserve.c-om
Contact: Mariam Nayiny Member: ATA

ISI, 3501 Market St., Philadelphia, PA 19104
Phone: (215)386-0100 Fax: (215)243-2208
e-mail: ellen.boyar@isinet.com Website: www.isinet.com
Contact: Ellen A. Boyar

ISI (Interpreting Services International Inc.), 6180 Laurel Canyon Blvd., #245, N. Hollywood, CA 91606 Phone: (818)753-9181 Fax: (818)753-9617 e-mail: info@isitrans.com
Contact: Cathi Rimalower

INAWORD, 811 W. 7th St., Ste. 204, Los Angeles, CA 90017
Phone: (213)688-6200 Fax: (800)805-7994
e-mail: info@inaword.net Website: www.inaword.net
Contact: Stella Fridman Member: ATA

Information Builders, Inc., Two Penn Plaza, New York, NY 10121
Phone: (212)736-4433 Fax: (212)967-6404 Website: www.ibi.com
Contact: Georgina Channer Member: ATA
Inline Translation Services, Inc., 100 W. Broadway, Suite 520, Glendale, CA 91210 Tel: (818)547-4995 Fax: (818)547-4013 www.inlinela.com

Inlingua Language & Intercultural Services, 15500 Wayzata Blvd., Suite 819, Wayzata, MN 55391 Phone: (952)929-9155 Fax: (952)929-0020
e-mail: info@inlingua-mn.com
Web site: www.inlingua.com/minneapolis.htm
Contact: Carolyn Leavenworth Meyers Member: ATA

Inlingua Translation Service, 95 Summit Ave., Summit, NJ 07901
Phone: (908)522-0622 Fax: (908)522-1433 e-mail: summit@inlingua.com
Contact: Erica Alicea
Founded 1968. Uses freelance translators and interpreters. Accepts unsolicited resumes. Resumes are kept on file for future jobs. Applicants must have computer, fax, and modem. Ten top languages are Spanish, French,Italian, Portuguese, Russian, German, Chinese, Japanese, Danish, and Polish. Also translates Korean and Romanian. Main areas are lincenses, birth certificates, and transcripts. Has pool of some 30 translators.Operates as language school with translation services.

Inlingua, Inc., 171 East Ridgewood Ave., Ridgewood, NJ 07450 Phone: (201)444-9500 Fax: (201)444-0116
E-mail: ridgewood@inlingua.com Web site: www.inlingua.com, 2nd website: www.inlinguanorthnj.com
Contact: Sandra Stern, Translation Services Coordinator
Founded 1977. Member: ATA. Accepts resumes from translators and interpreters.. Requires ATA accreditation. Main languages are Spanish, French, Portuguese, German, Italian, Chinese, Japanese, Russian, and Arabic. Also Hindi, Hungarian, Polish, Romanian and Dutch Czech, Serbo-Croation, Haitain-Creole, French Canadian. Mainly legal, pharmaceutical, financial. Additional subjects that we translate: websites, medical, cosmetics,electronics subjects, tape transcriptions. Uses some 150 translators.

Intermark Language Services Coporation, 2555 Cumberland Pkwy, Ste. 295, Atlanta, GA 30339 Phone: (770)444-3055 Fax: (770)444-3002
e-mail: info@intermark-languages.com
Website: www.intermark-languages.com
Contact: Tom West
Founded 1995. Member: ATA. Accepts resumes. Interested in translators with with a law or business degree. Translates mainly French, Spanish, German, Portuguese and Swedish. Emphasis is on Law and Business translation. Prospective translators *must* provide a sample translation in every combination they wish to translate; ATA accreditation is preferred.

InterNation, Inc., 299 Broadway, Ste. 1400, New York, NY 10007
Phone:(212)619-5545 Fax: (212) 619-5887
e-mail: info@internationinc. com Web site: www.internationinc.com
Founded 1990. Member: ATA, NYCT. Uses freelance translators and interpreters. Accepts unsolicited resumes. Prospective translators are invited to register themselves in the online database at www.internationinc.com. All applicable database forms must be filled out completely. Translates from and into all major languages, but predominantly works with Spanish, Portuguese, Italian, French, German, Japanese, Chinese, Russian, Korean, Arabic, Polish, Dutch and the Scandinavian languages. InterNation specializes in foreign language voiceovers and subtitling for corporate sales, image, training, safety, advertising videos and informercials as well as documentaries. Maintains a state-of-the-art digital recording studio in-house and has an online voice talent database with audio samples of 400 native language voice actors in over 40 languages. Employs narrators and character actors mainly in the New York City area. Also translates commercial, financial, legal, technical, scientific, medical, pharmaceutical, PR material and advertising copy. Maintains a pool of some 5,000 translators.

International Bureau of Translations, Inc, 10291 North Meridian St., Ste. 350, Indianapolis, IN 46290 Phone: (317)581-0060 Fax: (317) 581-1160 e-mail: ibtinc@ibtworld.com Web site: www.ibtworld.com
Contact: Demetria Gecewicz

Founded 1976. Member: OTIAQ, Indianapolis Chamber of Commerce, IUPUI, Indiana University. Translated 56 languages last year. Emphasis on Medical, Technical, Legal, Business and Personal Document translation. Prospective translators should be native speakers with access to a computer, modem, or fax. They should send a rate chart with their resume. Resumes are filed and the translator contacted as the need arises. In addition to translation and desktop publishing, the company does foreign videos, court interpretation, cultural presentations, and conducts language classes. They research the cultural acceptability of company names, product names and slogans, and are the official translators for the International Violin Competition. Maintains a pool of 150 translators.

International Communication by Design, Inc., 1726 North First St., Milwaukee, WI 53212 Phone:(414)265-2171 Fax:(212)265-2101
e-mail: info@icdtranslation.com Website: www.icdtranslation.com
Contact: Jennifer Marquardt, Office Manager
Founded 1991. Member: ATA, Wisconsin World Trade Center, STC. Translates mainly Italian, Spanish, French, German, Portuguese, Dutch, Arabic, Greek, Japanese and Chinese. Also Korean, Russian and Swedish. Emphasis on Technical Manuals, medical and Legal translation. Also website localization. Prospective translators should be native speakers and ATA accredited, with a specific area of expertise and must be able to send projects in via CompuServe. Resumes are filed and used as a resource for new translators, interpreters and typesetters. It is a translation service company with a wide range of clients, from large corporations to smaller cottage industries. The company is actively community involved. Maintains a pool of over 60 translators.

International Contact, Inc., 351 15th St., Oakland, CA 94612
Phone:(510)836-1180 Fax:(510)835-1314
e-mail: info@intlcontact.com Web site: www.intlcontact.com
Contact: Carla Itzkowich Member: ATA

International Effectiveness Centers, 360 Pine St., 3rd Floor, San Francisco, CA 94104 Phone:(415)788-4149 Fax:(415)788-4829
e-mail: iec@ie-center.com Website: www.ie-center.com
Contact: Taryk Rouchdy
Founded 1972. Member: ATA, NCTA. Translates Spanish, Chinese, Russian, and Portuguese, both from and into English, French and Vietnamese from English, and Japanese into English. Emphasis on Legal, Technical and Children's Book translation. Uses freelance translators and interpreters. Accept unsolicited resumes. Prospective translators must have at least 3 years experience and be tested unless they hold current ATA accreditation. Resumes are screened for qualifications and appropriate equipment and, if qualified, entered into a data base The company also does desktop publishing in all languages, as well as voiceovers, dubbing, promotional work, crosscultural and language training, and international business consulting. Maintains a pool of 7000 translators in its data base, 170 of whom are used regularly.

International Institute of Wisconsin, 1110 N. Old World 3rd St.,
Milwaukee, WI 83203 Phone: (414)225-6220 Fax: (414)225-6235
E-mail: iiw@exepc.com Web site: www.iiwisconsin.org
Contact: Alexander P. Durtka, Jr.

International Language Center, 1416 S. Big Bend Blvd., St. Louis, MO 63117 Phone: (314)647-8888 Fax: (314)647-8889
e-mail: ilc@ilcworldwide.com Website: www.ilcworldwide.com
Contact: Dede S. Brunetti

International Language Services, Inc., 5810 Baker Rd., Ste. 250,
Minnetonka, MN 55345 Phone: (612)672-9600 Fax: (612)672-9400
e-mail: info@ilstranslations.com Web site: www.ilstranslations.com
Contact: John Funk
Founded 1982. Translates Spanish, French, German, Italian, Portuguese, Japanese, Swedish, Dutch, Danish, Chinese and most other languages from English. Emphasis on Medical, Technical, Business and Legal translation.Prospecti-ve translators should have ATA accreditation, references, a minimum 3_5 years experience, and be native speakers of the target language. Resumes are read and filed by language. Their primary focus is on biomedical instrumentation. Maintains a pool of over 300 translators.

International Language Source, Inc., P.O. Box 338, Holland, OH 43528

Phone:(419)865-4374 Fax:(419)865-7725
e-mail: ilsource@adelphia.net Web site: www.ilsource.com
Contact: Ryan Stevens, Account Manager
Founded 1981. Member: TAITA, ATA. Uses freelance translators and interpreters. Accepts unsolicited resumes. Prospective translators should have a specialty area and the appropriate resources for projects in that area. Resumes are kept for 6 months in an active database. Contact may involve the paid translation of a short sample. Translates mainly French, Spanish, German, Italian and Chinese. Also Japanese, Portuguese and Hebrew. Emphasis on Glass Manufacturing and Retail, Automotive OEM and Aftermarket, Furniture, Legal, Industrial and Healthcare translation. In addition to translation services, the company provides video and interactive CD-ROM translation and voiceovers. The company is willing to consider internships and cooperative projects with other translation companies.

International Translating & Typesetting Company, 2730 Stemmons Freeway, Ste. 302, Dallas, TX 75207-2248
Phone:(214)630-0840
Fax: (214)630-6482 e-mail: 102462.161@compuserve.com
Contact: Javier Galdeano, Translation Manager
Founded 1969. Member: Greater Dallas Chamber of Commerce. Translates Spanish, Chinese, French, German, Arabic, Russian, Japanese, Portuguese, Korean, Vietnamese, Bulgarian, Cambodian, Serbo-Croatian, Czech, Danish, Dutch, Farsi, Finnish, Greek, Gujarati, Hebrew, Hindi, Indonesian, Kurdish, Latin, Norwegian, Polish, Punjabi, Romanian, Hungarian, Swedish, Tagalog, Thai, Turkish, Ukrainian, and Urdu, both from and into English. Emphasis on Petroleum, Electronics, Legal, Business, Advertising, Food Export, Chemical, and Civil Engineering translations. Prospective translators must accept work *only* from translation companies, and *never directly from end users*. They must have either IBM or Macintosh computers, a modem and a fax. Resumes are filed until a need arises. In addition to translation, the company also provides high-resolution typesetting in most languages. Maintains a pool of 200 translators.

International Translating Bureau, 16125 West 12 Mile Rd., Southfield, MI 48076-2912 Phone:(248)559-1677
Fax:(248)559-1679
e-mail: itbinc @itbtranslations.com
Contact: Mariano Pallarés, President
Founded 1977. Translates all language pairs. Emphasis on Automotive, Engineering, Machine Tool, Legal, Medical, and Public Relations translation. Prospective translators should be native speakers of the target language, have at least 5 years residence in the source-language country, and have academic training or on-job experience in a specialized field. Candidates must have computer, fax and current software, particularly WordPerfect and Word for Windows for PC or Mac. Re-sumes are scanned for qualifications in which the company is interested and filed for future use. The company offers accurate, native-like translations. Maintains a pool of 250 translators.

International Translation Service, P.O. Box 188331, Sacramento, CA 95818 Phone: (530)753-7482 Fax: (530)753-7482
E-mail: its@yolo.com Web site: www.geocities.com/its worldwide Contact: Garry Pratt
Founded 1970. Uses freelance translators and interpreters. Accepts unsolicited résumés. Prospective translators must be experienced in both their languages and specialty fields, and should possess a variety of means of communications, including phone, fax and e-mail; they should have computers with the appropriate language fonts. Résumés are filed by language. Translates mainly Scandinavian, German, Spanish, and French, and also Estonian, Irish, Italian, Russian, Finnish, Portuguese, Polish, Dutch and more. Translates mainly legal, scientific, commercial, and medical. Maintains a pool of some 80 translators.

International Translation Solutions (ITS), 990 Lumber Exchange Building, 10 South 5th Street, Minneapolis, MN 55402
Phone: (612) 339-4660 General e-mail: translate@intransol.com
Web site: http://www.intransol.com
Contact: Jason Wood, President and CEO
Founded 1989. Member: ATA, STC, NADTP, GBCC. ITS translates all major languages to and from English. Emphasis on technical, medical, legal, B to B, advertising and marketing materials. Prospective translators must have professional experience, be subject pecialized and be proficient with TRADOS or other CAT tools. Submit resumes to hr@intransol.com. ITS also provides multilingual typesetting and desktop publishing, graphic design, and audio and video production services. Maintains a pool of over 3,000 translators

Interpretations, Inc., 8665 W. 96th St., Ste. 201, Overland Park, KS 66212 Phone: (913)338-2434 Fax: (913)338-

2636
e-mail: interp@accessus.net Web site: www.exactwords.com
Contact: Gloria J. Donohue
Founded 1991. Member: ATA, MICATA, International Trade Club, International Relations Council, Women's Resources Network. Translates mainly Spanish, French, German, Portuguese, Russian, Dutch, Swedish and Italian. Also Danish, Finnish and Korean. Emphasis on Instructional and Technical, Legal, Advertising, Video Script and Educational Materials translation. Prospective translators must be native speakers and have five years' experience; ATA accreditation is a must. They should be up_to_date on hardware and software. Resumes are reviewed and filed; a select number are asked to provide samples of their work and/or translate short paragraphs. The company also provides interpreters for meetings, conferences and telephone conferencing. Maintains a pool of about 100 translators.

Interpreters' Index.com, P.O. Box 880872, San Diego, CA 92168
Phone: (619)850-8279 Fax: (619)334-9473
e-mail: info@interpretersindex.com
Website: www.interpretersindex.com
Contact: Juana Esther Blanco Member: ATA

InterSol, Inc.,Three Point Drive, Ste. 301, Brea, CA 92821
Phone: (714)671-9180 Fax: (714)671-9188
email: solutions@intersolinc.com Website: intersolinc.com
Contact: Susana Turbitt
Founded 1996. Member: ATA. Accepts resumes. Translates mainly French, Italian, German, Spanish, Portuguese, Japanese, Chinese, Danish, Swedish and Finnish. Mostly medical, electronics, computers, automotive, general business, travel and sports. Also software localization. Uses some 30 translators.

Interspeak Translations, Inc., 114 East 32nd St., New York, NY 10016 Phone: (212)679-4772 Fax: (212)679-5084
Contact: Silvia Zehn Member: ATA

IRU Language & Translation Services; 2909 Hillcroft Ave, Suite 538, Houston, TX, 77077; Phone: (713)266-0020
Fax: (713)266-1716; Web Page: www.iru_services.com e-mail: IRU2000@aol.com
Contact: Elke Krause;
Member: AATG, ATA, HITA, WBEA, GACC, Houston-Leipzig Sister City Association; Founded In 1989. The main languages that IRU translates are English, German, Spanish, French, Italian, Portuguese, Russian, Chinese, Korean and Indonesian. In addition they translate all other languages. The main subjects that they translate are legal, commercial, medical, technical and individual. IRU does use freelance and they do accept resumes. Once they have received a resume it is checked if all the information needed is enclosed and then it is added to a freelance database. They require translators to hold a translation degree or equivalent, interpreters to be court certified and teacher have to hold a degree for foreign language education. IRU Language & Translation Service is a well-established company, serving not only the greater Houston area, but also Baytown, Clear Lake, Orange and Freeport, Texas. Some more of their services include foreign language and ESL classes, private language tutoring with an emphasis on German, cross-cultural seminars and tour-guiding/chaperon services for visitors. They offer prompt, accurate and confidential services at competitive prices. They are also listed in the membership directory of the German American Chamber of Commerce. Their average work load per year is 1,500 translations per year. They have about 40 translators in their pool.

Italian Language Connection, Inc., 5225 Pooks Hill Rd. Ste. 504-S, Bethesda, MD 20814 Phone: (301)897-3728
Fax: (301)897-0872
E-mail: ItalTranslations@cs.com
Contact: Mark L. Pisoni
Founded 1991. Member: ATA. Accepts resumes. Only translates Italian. Main subjects are economics and finance, legal, business, marketing and medical. Uses 15 translators.

Iverson Language Associates, Inc., P.O. Box 511759, Milwaukee, WI 53203 Phone:(414)271-1144 ext.21
Fax:(414)271-0144
e-mail: steve@iversonlang.com Web site: www.iversonlang.com
Contact: Steven P. Iverson, President
Founded 1986. Member: ATA, STC, ASTD. Translates French, Spanish, Japanese, Italian, Arabic, Dutch, and Korean from English, and Portuguese, German,Polish, Chinese and all other languages, both from and into English.Emphasis on Technical

Manuals (large machines), Medical Operators' Manuals, Spec Sheets, Financial, Software, Advertising Packaging, and Legal translations. Prospective translators should have computer, fax, current software, e-mail. Must be native speakers in the target language, with a technical background and formal training in a specific field of concentration. Three years experience in the industry and good writing skills are preferred. Resumes are reviewed internally for the required criteria. Interested translators should complete the translator listing format http://www.iversonlang.com/contact/translator.asp. The company provides translation, interpretation, typesetting and video narration services in most major languages, as well as technical writing and illustration. Typesetting capabilities include Chinese and Japanese. Maintains a pool of 600 translators. Iverson Language Associates is a long-standing member of the American Translators Association, and a Founding Member of the Association of Language Companies.

-J-

JLS Language Corporation, 135 Willow Rd., Menlo Park, CA 94025 Phone: (415)321-9832 Fax: (415)329-9864
e-mail: info@jls.com Website: jls.com
Contact: Ms. Rikko Field, President
Founded 1977. Member: ATA. Translates Japanese, Chinese, German, French and Spanish, both from and into English, with other languages on request. Emphasis on High Tech translation. Prospective translators should have ATA accreditation, education and industry experience. Resumes are screened and filed, and the promising applicants tested. In addition to High Tech, the company also does Agribusiness, Biomedical and Patent translations, and provides Chinese and Japanese desktop publishing. Maintains a pool of "hundreds" of translators.

JTG, Inc., 105 N. Washington St., Ste. 300, Alexandria, VA 22314
Phone: (703)548-7570 Fax: (703)548-8223
e-mail: hr@jtg-inc.com Website: www.jtg-inc.com
Contact: Muriel Jérôme-O'Keeffe Member: ATA

Japan-America Management, Ltd., 2020 Hogback Rd., Ste. 17, Ann Arbor, MI 48105 Phone: (734)973-6101
Fax: (734)973-1847
e-mail: jamltdmi@ameritech.net
Contact: Landon Bartley Member: ATA

Japan Communication Consultants, 236 E. 47th St., 28th Fl., New York, NY 10017 Phone: (212)759-2033 Fax: (212)759-2149
e-mail: jcc@japancc.com Website: www.japancc.com
Contact: Mariko N. O'Hare Member: ATA

Japan Pacific Publications, Inc., 419 Occidental Ave. S. #509, Seattle, WA 98104 Phone: (206)622-7443 Fax: (206)621-1786
e-mail: andrewt@japanpacific.com Website: www.japanpacific.com
Contact: Andrew Taylor, President
Founded 1983. Member: ATA. Uses freelance translators and interpreters. Accepts unsolicited resumes. Resumes are entered into database and evaluated for skills and experience. Main language is Japanese. Also translates Chinese and Korean. Main areas are agriculture, business, bio-med, tourism, marketing, software and wood products. Maintains pool of 35 translators. Translation, DTP, pre-press, image setting.

John Benjamins Publishing Company, P.O. Box 27519, Philadelphia PA 19118 Phone: (215)836-1200 Fax: (215)836-1204
email: paul@ benjamins.com Contact: Paul Peranteau Member: ATA

Josef Silny & Associates, Inc., 7101 SW 102 Ave. Miami, FL 33173
Phone: (305) 273-1616 Fax: (305) 273-1984
e-mail: translation@jsilny.com Web site: www.jsilny.com
Contact: Marisa di Giovanni Member: ATA

JPMorgan Chase, 575 Washington Blvd., Jersey City, NJ 07310

Phone: (201)595-8000 e-mail: maria.ricci@jpmchase.com
Contact: Maria Ricci Member: ATA

-K-

The Kaiser Permanente Medical Group, 2241 Geary Blvd., Rm. 146, San Francisco, CA 94115 Phone: (415)202-2888 Fax: (415)202-3829
e-mail: gayle.tang@kp.org
Contact: Gayle Tang Member: ATA

Kern Corporation, 230 Park Ave., Ste. 925, New York, NY 10169 Phone: (212)953-2070
Contact: Eric Schloss

Korean Consulting & Translation Service; P.O. Box 154773, Irving, TX 75015 Phone: (972)255-4808 Fax: (214)853-5374
e-mail: sbammel@koreanconsulting.com
Web Page: www.koreanconsulting.c-om
Contact: Steven S. Bammel, President
Member: ATA, founded in 1999. The main language pair that this company translates is Korean & English. They do any type of subject and translate any type of Korean. They do use freelance and they do accept resumes. The resumes are filed once they are received. The requirements for applicants are that they are native English-speaking and translate Korean into English. They specialize in Korean and are always looking for translators that speak English natively for Korean to English translation. They do not need translators for English to Korean. They have many translators in their pool.

Kotru Direct, 1001 Avenue of the Americas, #1214, New York, NY 10018 Phone: (212)840-2300 Fax: (212)840-0004
e-mail: kdiny@aol.com
Contact: Ravi Kotru Member: ATA

Kramer Translation, 893 Massasso St., Merced, CA 95340
Phone: (209)385-0425 Fax: (209)385-3747
e-mail: keith@thailinguist.com Website: www.thailinguist.com
Contact: Keith & Marisa Ensminger Member: ATA

-L-

LRA Interpreters, Inc., 5150 Wilshire Blvd., #506, Los Angeles, CA 90036 Phone: (323)933-1006 Fax: (323)933-1153
e-mail: Ira@Lrausa.com Website: www. Lrausa.com
Contact: Abel Plockier Member: ATA
LangTech International, 5625 SW 170th Ave., Aloha, OR 97007
Phone: (503)646-2478 (to fax, call first)
e-mail: LangTech@oregonvos.net Website: www.oregonvos.com
Contact: Douglas Foran, Owner
Founded 1994. Member: Latin American Trade Council of Oregon. Does not accept resumes. Translates French (European and Canadian), Spanish, Italian and Portuguese (Brazilian and Lusitanian), both from and into English. Emphasis on Scientific/ Technical, Mechanical, Electronics, Automotive, Agricultural, Legal, Health Care/Medical and Employee Relations translation. Also interpretation and audio/visual narration.

Langua Translations, Inc., 30150 Telegraph Rd., Ste. 300, Bingham Farms, MI 48025 Phone: (248)645-6663 Fax: (248)645-5376
e-mail: rneely@languatutor.com Web site: www.languatutor.com
Contact: R.P. Neely

The Language Bank, 875 O'Farrell St., Ste. 105, San Francisco, CA 94109 Phone: (415)885-0827 Fax: (415)885-1304
Contact: Yen Nguyen-le, Manager
Founded 1987. Uses freelance translators and interpreters. Accepts unsolicited resumes. Resumes are screened for qualifications. Main languages are Spanish, Chinese, Vietnamese, Russian, Tagalog, Thai, Laotian, Cambodian, and French. Emphasis on social services. Pool of more than 100 translators.

The Language Bank, Inc., 1323 Lancaster Drive, Orlando, FL 32806
Phone: (407)894-3300 Fax: (407)894-7825
e-mail: info@languagae-bank.com Website: www.language-bank.com
Contact:Dennis Ray Merritt

The Language Center, 144 Tices Ln., E. Brunswick, NJ 08816
Phone: (732)613-4554 Fax: (732)238-7659
e-mail: tlc@thelanguagectr.com Website: www.thelanguagectr.com
Contact: Mary Majkowski
Founded 1967. Member:ATA, TCD. Accepts resumes. Translates all languages. Main areas are health care, pharmaceutical, medical equipment and technical manuals.

The Language Center Inc., 7 Gilliam Ln., Riverside, CT 06878-0520 Phone: (203)698-1907 Fax:(203)698-2043
e-mail: thelanguagecenterinc@worldnet.att.net
Contact: Siri Ostensen, President
Founded 1978. Translates (both ways) Spanish, French, German, Portuguese, all Scandinavian languages, Italian, Japanese, Chinese, and Russian. Emphasis on Technical, Commercial and Legal translation. Prospective translators should provide sample translations. Resumes are entered into a data base. Maintains a pool of 30 translators.

Language Company Translations, L.C., P.O. Box 721507, Norman, OK 73070 Phone: (405)321-5380 Fax: (405)366-7242
e-mail: 71564.2044 @compuserve.com
Contact: Nancy T. Hancock, Director
Founded 1982. Member: ATA. Translates all languages in all technical disciplines. Potential translators should have experience and total fluency. They should be able to produce translations on disk if needed. Resumes are acknowledged and filed. Freelancers are contacted when appropriate work comes in. The company can take materials through final printing, working with commercial printers and supervising production. They also accept resumes from freelance interpreters. Maintains a pool of 40 translators.

Language Direct, 9801 Westheimber, Ste. 302, Houston, TX 77042
Phone: (713)917-6870 Fax: (713)917-6806
e-mail: jylan@langdirect.com Website: www.langdirect.com
Contact: Jylan Saleh-Maloy Member: ATA

Language Dynamics, 931 Howe Ave. #107, Sacramento, CA 95825
Phone: (916)920-4062 Fax: (916)920-3594 e-mail: juanvallej@aol.com Contact: Rhody Vallejo
Founded 1979. Translates Spanish, French, Italian, Russian and Chinese and Hmong. All subject areas are translated. Prospective translators should be of high quality. Resumes are read and kept on file for contact on an as-needed basis.

The Language Exchange, Inc., P.O. Box 750, Burlington, WA 98233
Phone: (360)755-9910 Fax: (360)755-9919
e-mail: Langex@langex.com Website: www.langex.com
Contact: Jaye Stover Member: ATA

The Language Group, 2421 Bowland Pkwy, Ste. 104-B, Virginia Beach, VA 23454 Phone: (757)431-9004 Fax: (75)431-0447
e-mail: info@thelanguagegroup.com
Website: www.thelanguagegroup.com
Contact: Evelin Schuyler

Founded 1999. Member: ATA, TCD, NAJIT, ASTM, NCMA, ABWA, NCATA. Accepts resumes via e-mail. Translates mainly Chinese, Croatian, Czech, French, German, Italian, Japanese, Norwegian, Polish and Spanish. Mostly technical, engineering, finance and marketing. Uses some 150 translators.

Language Innovations, LLC, 1725 I St. NW, Ste. 300, Washington, DC 20006 Phone: (202) 349-4180 Fax: (202) 349-4182
e-mail: Translate@languageinnovations.com
Web site: www.Languageinnovations.com
Contact: Brian S. Friedman Member: ATA

Language Intelligence, Ltd., 16 North Goodman Street, Rochester, NY 14607 Tel. (585) 244-5578 Website: www.languageintelligence.com

Language Learning Enterprises (LLE), 1627 K Street NW, Suite 610,
Washington, DC 20006 Phone: Phone: 888.464.8553 x224 Fax: (202)785-5584 e-mail: hlacy@lle-inc.com Web site: www.lle-inc.com
Contact: Heidi G. Lacy
Founded 1979. Resumes are carefully screened; prospective translators and interpreters are then interviewed over phone, references checked, and then tried out on a small project first. Requires minimum 2 years experience. Translates all languages "from Arabic to Zulu," both from and into English. Emphasis in all technical areas, with a specialty in medical interpretation. The company is the creator of LLE-Link, a 24-hour-a-day, 365 days-a-year telephone interpretation service using conference calling. Maintains a pool of 12 full-time people with 1000 linguists on call. With its rapidly growing client base LLE(tm) has opportunities for highly qualified linguists in each of its business units: instruction, face-to-face interpretation, translation, and telephonic interpretation. If you are interested in working with LLE on a freelance basis, please go to our website at www.lle-inc.com for information on positions and languages we are currently recruiting for. Resumes may be submitted under the Jobs tab/Contract Positions.

Language Line Services, 1 Lower Ragdale Drive, Bldg. 2, Monterey, CA 93940 Phone: (831)648-5819 Fax: (800)752-0073
e-mail: danyune@languageline.com Website: www.languageline.com
Contact: Daynune Geertsen

Language Link Corporation, 3100 Broadway, Ste. 110, Kansas City, MO 64111 Phone: (816)753-3122 Fax: (816)753-3262
e-mail: translation@languagelinkcorp.com
Website: www.languagelinkcorp.com
Contact: Carrie Hewitt
Founded 1989. Member: ATA, ACTFL, MICATA, ASTM. Uses freelance translators and interpreters. Accpts unsolicited resumes. Main languages are Spanish, Russian, Chinese, Vietnamese, Arabic, Croatian, Japanese, Somali, Korean and Farsi.

Language Matters, 1445 Pearl St., Ste. 215, Boulder, CO 80302
Phone: (303)442-3471 Fax: (303)442-5805
e-mail: info@languagematters.com Web site: www.languagematters.com
Contact: Rosangela Fiori
Founded 1992. Member: ATA. TCA. Translates medical, technical, legal documents, patents, marketing material, in all languages. Has a pool of over 800 translators.

The Language Network, Inc., 5151 Belt Line Rd., Ste. 452, Dallas, TX, 75240 Phone: (972)960-9980 Fax: (972)960-9904
e-mail: Language@flash.net
Contact: Christine J. Finck Member: ATA
LanguageOne, 2305 Calvert St. NW, Washington DC 20008
Phone: (202)328-0099 Fax: (202)328-1610
e-mail: translations@languageone.com Web site: www.languageone.com
Contact: Lorraine B. Smith Member: ATA

Language Plus,4110 Rio Bravo, Ste. 202, El Paso TX 79902
Phone: (915)544-8600 Fax: (915) 544-8640

e-mail: langplus@aol.com Website: www.languageplus.com
Contact: Connie Gyenis Member: ATA

The Language Service, Inc., 806 Main St., Poughkeepsie, NY 12603 Phone: (845)473-4303 Fax:(845)473-4467
e-mail: info@tlsmedtrans.com Website: www.tlsmedtrans.com
Contact: Jeanne De Tar
Founded 1950. Member: ATA, FIT. Uses freelance translators and interpreters. Accepts unsolicited resumes. Qualified linguists are filed for future reference. Main languages (both ways) are German, French, Spanish, Italian, and Portuguese, and Japanese into English. Also Dutch and Swedish into English. Main areas are medical, pharmaceutical, chemical and environmental. Does about one million words a year. Works with various translators, but mainly with a steady group of about ten. Owner is co-founder and former president of the ATA.

Language Services Associates, Inc., 607 North Easton, C, PO Box 205,
Willow Grove, PA 19090 Phone: (215)657-6571 Fax: (215)659-7210
e-mail: lschriver@lsaweb.com Web site: www.lsaweb.com
Contact: Laura T. Schriver
Founded 1991. Member: ATA, DVTA, NAJIT, CHICATA, NOTA. Uses freelance translators and interpreters. Accepts unsolicited resumes. Resumes are acknowledged, kept, and used when needed. Applicants should have ATA accreditation. Main languages (both ways) are Spanish, German, French, Chinese, Japanese, Korean, Vietnamese, Russian, Hindi, Italian, and Tamil. Also Cambodian, Arabic, Farsi, Hebrew, Dutch and more. Emphasis on Legal and Medical translation. Maintains a pool of 745 translators. We have 4,000 subcontractors.

Language Services Consultants, Inc., P.O. Box 412, Ardmore, PA 19003 Phone: (610)617-8962 Fax: (610)617-9108
e-mail: ruth.karpeles@verizon.net
Contact: Ruth S. Karpeles
Founded 1992. Member: ATA, DVTA. Accepts resumes. Translates mainly Spanish, Vietnamese, Cambodian, Russian, Chinese, Korean, French, Portuguese, Greek, Dutch and Albanian. Mostly medical, education and social services. Uses some 150 translators.

Language Source; P.O. Box 13114, Wauwatosa, WI 53213
Phone: (414)607-8766 Fax: (414)607-8767
e-mail: langsource@aol.com
Contact: Elizabeth Brudele-Baran

Language Training Centre, Inc.: 5750 Castle Creek Parkway, Suite 387, Indianapolis, IN, 46250: Phone: (317)578-4577 Fax: (317)578-1673; Web Page: www.languagetrainingcenter.com
e-mail: mgeorge@languagetrainingcenter.com
Contact: David Hunt
Member: ATA, TESOL, IFLTA, INTESOL. Founded in 1993. The Language Training Center translates English to Spanish, Spanish to English, English to German, German to English, French to English, English to French, Chinese to English, English to Chinese, Japanese to English and English to Japanese. The main subjects that they translate are handbooks and legal contracts. They specialize in corporate documents. They do use freelance and they also accept resumes. One they receive them, they file them by language and special area. They require that you have experience. They are a full-service language center that teaches, translates and acclimates clients in the languages of the world. A leader in the translation field for over 20 years. They have 40 translators in their pool.

The Languageworks, Inc.,1123 Broadway, Ste. 201, New York, NY 10010 Phone: (212)447-6060 Fax: (212)447-6257
e-mail: talwood@languageworks.com Website: www.anguageworks.com
Contact: Thomas Alwood Member: ATA

Latin American Translators Network, Inc.,1970 Cliff Valley Way NE, Ste.
208, Atlanta, GA 30329
Phone: (404)634-2635 Fax: (404)634-9683
e-mail: latn@latninc.com Web site: www.latninc.com

Contact: Alicia V. Mitchell Chacón   Member: ATA

Lazar & Associates, 2444 Wilshire Blvd., Ste. 411, Santa Monica, CA 90403   Phone: (310)453-3302   Fax: (310)453-6002
e-mail: languages@lazar.com   Website: www.lazar.com
Contact: Elaine Lazar
Founded 1995. Member: ATA. Accepts resumes occasionally.Ma-in languages are Spanish, French, Chinese, Japanese and German. Mostly medical, pharmaceutical, IT, business, legal and technical equipment manuals. Has a pool of over 3000 translators.

Le Chateau Enterprises Group, Inc., 10051 McGregor Blvd, Ste. 105
Ft. Myers, FL 33919   Phone: (941)574-5600   Fax: (941)574-3273
e-mail: info@letspeak.com   Website: www.letspeak.com
Contact: Allendy Doxy   Member: ATA

Legal Interpreting Services, Inc., 21-50 44th Drive, Long Island City, NY 11101   Phone: (718)786-7890   Fax: (718)786-8034
e-mail: accutran@aol.com   Website: www.accuratetranslation.com
Contact: Alexandra Gerenburg   Member: ATA

Leggett & Platt, Inc., PO Box 757, Carthage, MO 64836
Phone: (417)358-8131   Fax: (417)359-5767
Website: www.leggett.com
Contact: Tomasita Hernandez Member: ATA

Lemoine International, Inc., 339 E. 600 N., American Fork, UT 84003
Phone: (801)763-5644   Fax: (801)763-5660 e-mail: brigittem @lemoine-international.com   Website: www.lemoine-international.com
Contact: Brigitte L. Morales   Member: ATA

Letspeak, Inc., Phone: (239) 274-5700   Fax: (239) 274-9709
e-mail: info@letspeak.com   Web site: www.letspeak.com
Contact: Allendy Doxy   Member: ATA

Liaison Language Center, 3500 Oak Lawn Ave., Ste. 110 LB 32, Dallas, TX 75219   Phone: (214)528-2731 Fax: (214)522-7167
e-mail: liaison@liaisonlanguage.com Website: www.liaisonlanguage.com
Contact: Gerda Stendell
Founded 1978. Member: ATA. Uses freelance translators and interpreters.. Translated mainly Spanish, Portuguese, German, French, Italian, Korean, Japanese, Chinese (Madarin and Cantonese) and Vietnamese. Mostly legal, oil, insurance. Has a pool of around 100 active translators and interpreters.

Liaison Multilingual Services, 3261 Cherryridge Rd., Englewood, CO 80110   Phone: (303)762-0997   Fax: (303)762-0999
e-mail: liaison@ecentral.com   Website:eMultilingual.com
Contact: Suzanne Robinson   Member: ATA

Lingo Systems, 15115 SW Sequoia Pkwy, #200, Portland, OR 97224
Phone: (503)419-4856   Fax: (503)419-4876
e-mail: info@lingosys.com   Website: www.lingosys.com
Contact: Maja Bailey
Founded 1992. Member: ATA. LISA, SAO. Accepts resumes. Requires 5 years experience and knowledge of CAT tools. Translates mainly French, German, Italian, Spanish, Japanese, Chinese, Korean, Dutch, Scandinavian languages and Eastern European. Also Asian. Mostly IT, computer hardware and software, websites, marketing and training materials. Maintains some 500 translators.

Lingua Communications Translation Services, 9321 Lavergne, Ste. 101, Skokie, IL 60077 Phone:(847)673-1607 Fax:(847)673-1669
Contact: Alex Babich, General Partner
Founded 1989. Member: ATA. Translates Russian and German, both from and into English, and Polish, Ukrainian, Latvian, French and Japanese into English. Emphasis on all subject translations. Prospective translators should possess experience, speed and quality at market-level pricing. Resumes are filed against need; translators are then contacted. **Lingua** maintains a pool of 250 translators.

Lingua Language Center, 1730 Main St., Ste. 226, Weston, FL 33326
Phone: (954)349-8088 Fax: (954)349-1799
e-mail: info@linguaschool.com Website:linguaschool.com
Contact: Andreina Ojeda Member: ATA

Lingualink Incorporated, 5777 W. Century Blvd., Ste. 1000, Los Angeles, CA 90045 Phone: (310)642-2920 Fax: (310)642-2945
e-mail: lingualinkinfo@ccscom.com Website:
www.ccscom.com/ lingualink
Contact: Keir Milan Member: ATA

Linguistic Consulting Enterprises, Inc., 251 W. Marlboro Rd., Oro Valley, AZ 85737 Phone: (520)297-6420 Fax: (520)297-2629
e-mail: LCEAZ1@aol.com
Contact: Maria Elisa Rivera

Linguistic Systems, Inc., P.O. Box 390031, 130 Bishop Allen Drive, Cambridge, MA 02139 Phone:(617)864-3900 Fax:(617)864-5186
e-mail: info@linguist.com Website: www.linguist.com
Contact: Martin Roberts, President
Founded 1967. Member: ATA. Uses freelance translators and interpreters. Applicants should have relevant education and experience. Translates mainly Spanish, French, German, Italian, Portuguese, Chinese, Japanese, Korean, Greek and Swedish. Mostly legal, medical and commercial. The company provides translation, interpretation, voiceover, and machine translation editing services. Maintains a pool of about 4000 translators.

Lionbridge, 301 Mission St., Ste. 350, San Francisco, CA 94105
Phone: (415)546-6885 Fax: (415)495-4926
Website: www.lionbridge.com
Contact: Allison Smith Member: ATA

Littleton & Associates, Inc., 6006 W. 159th St., Bldg. C, Ste. 2 East, Oak Forest, IL 60452 Phone: (708)429-3336 Fax: (708)429-3414
e-mail: lcs62258@aol.com
Contact: James M. Littleton Member: ATA

Lucent Technologies Global Translation, 2400 Reynolda Rd., Winston-Salem, NC 27106 Phone: (336)727-6218 Fax: (336)727-3221
e-mail: mrhodes@lucent.com Website: www.lucent.com/translations
Contact: Marta C. Rhodes Member: ATA

LUZ, 202 Potero Ave., San Francisco, CA 941103
Phone: (415)241-0520 Fax: (415)241-0504 Website: www.luz.com
Contact: Sanford Wright Member: ATA

-M-

M² Limited, 9210 Wightman Rd., Montgomery Village, MD 20886
Phone: (301)977-4281 Fax: (301)926-5046
e-mail: info@m2ltd.com Website: www.m2ltd.com
Contact: Audrey Moyer
Founded 1979. Member: ATA, STC, World Trade Center. Accepts unsolicted resumes. Applicants should be native speakers of the target language, with 2-5 years minimum experience. Translates mainly Spanish, French, Italian, German, Porutguese, Chinese, Japanese, Korean, Dutch and Russian. Emphasis on Computer Software, technical documentation, web pages and localization. The company does not provide interpreting services. The company has become a specialist in the localization of computer-based equipment and software, and is heavily involved in software development, training, and adaptation of English-language materials for use in other cultures and languages. In addition to translators, the company is interested in proofreaders and editors in all languages. Maintains a pool of 2,000 translators.

M/C International, 7207 Chagrin Rd., Unit #4, Chagrin Falls, OH 44023 Phone:(440)247-2277 Fax:(440)247-0453
e-mail: mcinternational @stratos.net
Contact: Mrs. E. Marchbank
Founded 1969. Translates Spanish, French, German, Italian, Portuguese, Russian, Accepts resumes. Translates mainly French, German, Dutch, Italian, Portuguese, Spanish, Arabic, Chinese and Japanese. Emphasis on technical, legal and commercial translation. The company is a full-service translation bureau. It also provides 4-color printing services, desktop publishing and typesetting, both to clients and to other translators. In addition, they operate their own travel agency. Maintains a pool of 50 translators.

MGE Lingual Services, 136 36th St. Drive SE, Ste. A-4, Cedar Rapids, IA 52403 Phone: (319)366-1038 Fax: (319)366-1047
e-mail: mike@mge-lingual.com Website: www.mge-lingual.com
Contact: Michael G. Elliff Member: ATA

MTS Multinational Translating Service, 928 Connetquot Ave., Central Islip, NY 11722 Phone: (631)581-8956 Toll Free: (800)864-5069, 70 & 71 Fax: (631)224-9435 e-mail: info@mtsinc.us Web site: www.mtsinc.us
Contact: Lisa Alesci Member: ATA

Madison Linguistic Services, 301 Keezell Hall, MSC 1802, Harrisonbur-g, VA 22807 Phone: (540)568-3512 Fax: (540)568-6904
e-mail: rethorcx@jmu.edu Web site: jmu.edu/forlang/trans
Contact: Christophe Rethore Member: ATA

MAGNUS International Trade Services Corp., 1313 North Grand Ave., #280, Walnut, CA 91789 Phone: (909) 595-8488 Fax: (909)598-5852 Website: www.manuscorp.com
Contact: Nora K. Schwartz
Member: ATA, PCLA, HPRMA, LA-FTA, NOTA. Translates Spanish, Chinese, Vietnamese, Tagalog, Korean, Portuguese, French, German and Russian from English, as well as Hindi, Farsi, Armenian, Hmong, Lao and Arabic. Emphasis on Healthcare, Medical, Pharmaceutical, Engineering, Legal, Technical, and General Business translation. Prospective translators should have certification (if possible), appropriate education and field experience. Resumes are responded to with a standard contract, and contractors called as needed.

Marion J. Rosley Secretarial, Transcription & Translation Services, 41 Topland Rd., Hartsdale, NY 10530 Phone: (914)682-9718 Fax: (914) 761-1384 e-mail: mrosley@rosley.com Website: rosley.com
Contact: Marion J. Rosley, President
Founded 1977. Member: NYCT, World Trade Council of Westchester, Westchester County Assn., Westconn. Translates all languages. Emphasis on Legal, Medical, Technical, and Business Correspondence translation. The company provides interpreting services in all languages. Prospective translators should have excellent skills and reasonable prices. Resumes are filed for future use. Maintains a pool of "hundreds" of translators. The company provides secretarial, transcription and translation services, including cassette transcription of conferences.

MasterWord Services, Inc., 303 Stafford, Ste. 204, Houston, TX 77079 Phone:(713)589-0810 Fax:(713)589-

1104
e-mail: masterword@ masterword.com Website: www.masterword.com
Contact: Mila Green
Founded 1993. Member: ATA, Azeri Translators Association. Translates Russian, Spanish, Azeri, Brazilian Portuguese and Chinese, both from and into English. Emphasis on Oil and Gas, Drilling, Legal and Medical translations. Prospective translators should have significant translation experience, especially in the oil/gas or legal areas, and show quality performance on the company's certification exam. All resumes are acknowledged, and after prescreening, ap-proximately 10% are sent an in-house certification exam, primarily in Russian or Spanish. They are then phoned for an in-person interview. The company also has offices in Baku, Azerbaijan, and is involved in ongoing projects in Europe, South America and Asia.They also have affiliates in Washington, DC, New York, London and Moscow. Maintains a pool of 35-50 translators.

Masua Funai Eifert & Mitchell, Ltd., 1701 Golf Rd., Ste. 800, Rolling Meadows, IL 60008
Contact: Yasuyo Wakuta Member: ATA

McNeil Multilingual, 6564 Loisdale Court, Ste. 800, Springfield, VA 22150 Phone: (703)921-1600 Fax: (703)921-1610
e-mail: rnelson@mcneilmultic.com Website: www.mcneilmultic.com
Contact: George E. Washington, Jr. Member: ATA

ME Sharp, Inc., Publisher, 80 Business Park Drive, Armonk, NY 10504 Phone: (914)273-1800 Fax: (914)273-2106
e-mail: journals@mesharpe.com Website: www.mesharpe.com
Contact: C. P. Chetti Member: ATA

Mena's International Corp., 2548 Nicollet Ave., S., Minneapolis, MN 55404 Phone: (612)872-8392 Fax: (612)872-7861
e-mail: menastrave@aol.com
Contact: Maximo Mena Member: ATA

Mercury Marine, W6250 West Pioneer Rd., Fond du Lac, WI 54936 Phone: (414)929-5299 Fax: (414)231-8916
e-mail: Gary_Fenrich@mercmarine.com
Contact: Gary Fenrich Member: ATA

Metropolitan Interpreters and Translators Worldwide, Inc., 110 E. 42nd St., Ste. 802, New York, NY 10017
Phone: (212)986-5050 Fax: (212)983-5998 e-mail: metlang@aol.com Website: metlang.com
Contact: Joe Citrano
Founded 1990. Member: ATA, NYCT, NAJIT, IAFL. Uses freelance translators and interpreters. Accept unsolicited resumes. Resumes are entered in a database. Translates all languages. Emphasis on legal and law enforcement. About 500 in translator pool.

Merrill Translations; 225 Verick St., New York, NY 10014
Phone: (212)367-5970 Fax: (212)367-5969
e-mail: translations@merrillcorp.com Web Page: www.merrillcorp.com resumes: resumestrans@merrillcorp.com
Contact: Thomas Alwood
Member: ATA, ITI; Founded in 1998. Translates mainly English to Spanish, French, German, Italian, Portuguese, Japanese, Chinese, Korean, Swedish and Russian. They work in 64 different languages. Subjects translated are legal, financial, advertising, medical and technical. They primarily focus on document translation. They also do some interpreting. Merrill Translations does use freeland and they do accept resumes. The resumes are all reviewed and qualified candidates are contacted to continue the registration process. They are mainly interested in applicants with college education and experience in legal and financial work. Merrill Translations is the foreign language services division of Merrill Corporation, a diversified document services company. They have offices in over 50 locations in the US and Europe. Merrill Translations production offices, located in New York and London, offer a full range of foreign language services, focused on the financial, life sciences, advertising, advertising and legal industries as well as general corporate clients. In addition to translation, localization, interpreting and foreign language typesetting services they also coordinate services with Merrill Corporation's other document management solutions, including printing and distribution anywhere in the world. They have more than 2,000 translators in their pool.

Monti Interpreting & Translation Services, Inc., 101 S. Wymore Rd., Ste. 315, Altamonte Springs, FL 32714
Phone: (888)686-6007 Fax: (407)830-1860
e-mail: montitrans@yahoo.c-om Website: www.montitrans.c-om
Contact: Caroline Montalvo
Founded 1997. Member: ATA. Accepts resumes. Translates mainly Spanish, Creole, Vietnamese, French, Chinese, Arabic, Bosnian, Croatian, Cambodian, Portuguese and Turkish. Mostly medical, legal and corporate literature. Maintains a pool of 50 translators and 500 interpreters.

Morales Dimmick Translation Service, Inc., 1409 W S Slope, Emmett, ID 83617 Phone: (208)365-2622
e-mail: projects@mdtranslation.com Website: www.mdtranslation.com
Contact: Chris
Founded 1989. Member: ATA, IMA. Accepts resumes. Main languages are Spanish, Bosnian, Laotian, Vietnamese and Arabic. Mostly business, medical, legal and insurance. Maintains a pool of 65 translators.

Morgan Guaranty Trust Company, 60 Wall St., 34th Fl., New York, NY 10260 Phone: (212)552-5474 Fax: (212)552-8128
e-mail: lisa.gallo@chase.com
Contact: Lisa Gallo Member: ATA

The Multi-Lingual Group, 8 Faneuil Hall Marketplace, 3rd Fl., Boston, MA 02109 Phone: (617)973-5077 Fax: (617)787-3124
e-mail: multilg @aol.com
Contact: Felice Bezri
Founded 1994. Member: NAJIT. Translates Spanish, French, Italian and Portuguese, both from and into English, German, Swedish, Arabic and Cantonese into English, and Japanese and Mandarin from English. Emphasis on Medical Studies, Computer, Legal and Military translation. Prospective translators should be ATA members with extensive experience in written translation and specialization in one or more fields. Resumes are reviewed on a regular basis with a particular eye to the experience of the candidate. The company also provides simultaneous interpreting in over 50 languages and offers cross-cultural consulting. Maintains a pool of approximately 320 translators.

MultiLing International, Inc., 55 N. University Ave., Ste. 225, Provo, UT 84601 Phone:(801)377-2000 Fax:(801)377-7085
e-mail: translation@ultiling.com Website: www.multiling.com
Contact: Michael V. Sneddon
Founded 1988. Member: ATA, STC, Utah Information Technology Assn., Software Publishers Assn. Translates Spanish, French, Dutch, German, Italian, Japanese, Chinese, Swedish, Danish, Russian and all other languages, both from and into English. Emphasis on High-Tech Product Localization and the translation of related documents. Prospective translators must be native speakers of the target language with an undergraduate and, preferably, a master's degree, with an area of special expertise. They should have a good software and desktop publishing background and several years of experience and accreditation. The company works in all areas of translation, including working with translation software and aiding in its further development. Maintains a pool of1400 translators.

Multilingual Communications Corporation, P.O. Box 7164 Pittsburgh, PA 15213 Phone:(412)621-7450 Fax:(412)621-0522
e-mail: mktg @mccworld.com Website: www.mccworld.com
Contact: Cathy Rosenthal
Founded 1977. Member: ATA. Translates Spanish, French, German, Chinese, Japanese, Russian, Italian, Korean, Portuguese, Arabic and Korean, both from and into English. Emphasis on Technical, Legal and Medical Software translation. Prospective translators should possess native proficiency and background in a specialized field. Resumes are evaluated on receipt, then contacted and tested. Their services include software localization. Maintains a pool of 500 translators.

Multilingual Solutions, 22 W. Jefferson St., Ste. 402, Rockville, MD 20850 Phone: (301)424-7444 Fax: (301)424-7331
e-mail: laurie@msolutions.com Website: www.mlsolutions.com
Contact: laurie Campos Sandler Member: ATA

Multilingual Translations, Inc., 1510 Front St., Ste. 200, San Diego, CA 92101 Phone: (619)295-2682 Fax: (619)295-2984
e-mail: mcardenas@multitrans.com Website: www.multitrans.com
Contact: Michael Ramírez Cardénas, J.D. Member: ATA

Multilingual Word, 2631 Salem Ave. S, Minneapolis, MN 55416
Phone: (952)929-4203 Fax: (952)922-4344 e-mail: mlivon@aol.com
Contact: Michelle Livon Member: ATA

-N-

NCS Enterprises, Inc., 1222 Hope Hollow Rd., Carnegie, PA 15106 Phone: (412)278-4590 Fax: (412)278-4595
e-mail: cnagy@ncs-pubs.com Web site: www.ncs-pubs.com
Contact: Charlene Nagy
Founded 1992. Member: ATA. Uses freelance translators and interpreters. Accepts unsolicited resumes. All resume are reviewed and entered into database.

NiS International Services, 1913 Dundin Drive, Old Hickory, TN 37138 Phone:(615)758-6459 Fax:(615)758-6459
e-mail: Nisintl@aol.com Website: www.nisintl.com
Contact: Bill Hollaway
Founded 1987. Member: ATA. Translates Japanese, Korean, Chinese, Spanish, German, French, Italian, and Russian, both from and into English. Emphasis on Automotive, Electrical Engineering, Mechanical Engineering, Patent, Contract and Agreement, and other technical document translation. Prospective translators should send a sample of previous work. A specialized technical background is a plus. Resumes are saved in a database, and the translator contacted when a job is available. The company provides translation, interpretation, and teaching services. Maintains a pool of 600-700 translators.

N.O.W. Translations, 23054 Covello St., West Hills, CA 91307
Phone: (818)716-9112 Fax: (818)888-6962
e-mail: thomasc@nowtranslations.com
Website: www.nowtranslations.com
Contact: Thomas Clement Member: ATA

Narragansett Translations & Interpreting, Inc., 32 East Ave., Pawtucket, RI 02860 Phone: (401)722-1222 e-mail: morra_red@ids.net
Contact: Mike D. Chea Member: ATA

Nelles Translations, 104 South Michigan Ave., Room 1004, Chicago, IL 60603 Phone: (312)236-2788 Fax: (312)236-0717
Contact: Lawrence R. Kramer, Product Manager
Founded 1956. Member: ATA, Chicagoland Chamber of Commerce, Chicago Convention & Tourism Bureau. Translates Spanish, French, German, Italian, Portuguese, Polish, Russian, Japanese, Chinese, Korean and numerous other languages, both from and into English. Emphasis on Patent, Print Advertising, Legal, Brochures and Manuals, and Personal Document translation. Prospective translators should be experienced with references. They may be tested at the discretion of the company. Resumes are acknowledged and promising applicants are contacted. The company provides interpreting and voiceover services in addition to translation.

NetworkOmni Multilingual Communications, 1329 E. Thousand Oaks Blvd., 2nd Fl., Thousand Oaks, CA 91362
Phone: (805)379-1090
Fax: (805)379-2467 e-mail: omni@networkomni.com
Website: www.networkomni.com Contact: Irena Stone
Founded 1978. Member: ATA. Translates French, Italian, German, Spanish, Chinese, Korean and Japanese, both from and into English. Handles a variety of subjects. Excellence is the primary requirement for prospective translators. Resumes are followed up with a call and a registration skills packet. The company has five separate divisions: Interpretations, Translations, Teleinterpretations, Transcriptions and Videointerpretations and all are supported by Omni's Quality Control staff. Maintains a pool of 800 translators.

New England Translations, 59 Temple Place, Ste. 204, Boston, MA 02111 Phone: (617)426-4868 Fax: (617)695-

9349
e-mail: netranslations@langcenters.com
Website: newenglandtranslations.com
Contact: Ken Krall, Director
Founded 1986. Member: ATA. Uses freelance translators and interpreters. Accepts unsolicited resumes. Resumes are entered into databse by language and specialty. Requires experience. Translates Spanish, Portuguese, French, German, Italian, Chinese, Vietnamese, Korean and Japanese from English, and Spanish into English. Emphasis on Finance, Legal, Marketing, Industrial and Medical translation, with specialties in Financial and Medical. Prospective translators should have PC or Mac, fax, modem. A small sample translation should be sent with the resume. Resumes are entered into a database by language pair and technical specialty after qualifications are reviewed. The company is dedicated to the translation of all major languages. It also provides instruction in Spanish only. Maintains a pool of 800 translators.

Newtype, Inc., 1259 Route 46 E, Bldg. 1, 2nd Fl., Parsippany, NJ 07054
Phone: (973)263-5000 Fax: (973)263-5522
e-mail: tbias@newtypeinc.com Website: www.newtypeinc.com
Contact: Thomas Bias Member: ATA

Northwest Translations, 2636 E. Greentree Ct., Eagle, ID 83616
Phone: (208)938-5005 Fax: (509)351-7529
e-mail: sales@nwtranslations.com Website: www.nwtranslations.com
Contact: Jeff Scott Allen Member: ATA

NovaTrans Enterprises, Inc., P.O. Box 3903, Boca Raton, FL 33427
Phone: (561)368-0865 Fax: (954)255-5161
e-mail: mail@novatrans.net
Contact: Janet T. Aliaga Member: ATA

-O-

Okada & Sellin Translations, LLC 1950 Addison St., Ste. 101, Berkeley CA 94704 Phone: (510)843-5600 Fax: (510)843-5603
e-mail: okada@ostrans.com Website: www.ostrans.com
Contact: Robert G. Sellin
Founded 1996. Member: ATA, STC. Uses freelance translators and interpreters. Accepts unsolicited resumes. Resumes are reviewed and graded. Work sample are welcome. Requires e-mail. Translates all major Asian and European languages. Focuses on patents and other technical, scientific, pharma/biotech, and legal subjects.

Omega Translation Service, 331-A North Brand Blvd., Glendale, CA 91203 Phone: (818)545-3626 Fax: (818)545-0364
e-mail: omega2@earthlink.net
Contact: Jill Massehian
Founded 1985. Translates Spanish, French, German, Japanese, Italian and Chinese, both from and into English. Emphasis on Scientific, Businese and Computer translation. Prospective translators should have good references. Resumes are filed by specialty and language. The company needs qualified translators and interpreters with various specialty fields. Maintains a pool of 1000 translators.

OmniLingua, Inc., 306 6th Ave. SE, Cedar Rapids, IA 52401 Phone: (319)365-8565 Fax:(319)365-7893 e-mail: omnilingua @omnilingua .com Website: www.omnilingua.com
Contact: Elizabeth Miller
Founded 1980. Member: ATA, MACHETE, NCATA. Accepts resumes. Translates mainly French, Italian, German, Spanish, Portuguese, Chinese, japanese, Korean, Greek and Russian. Mostly automotive, heavy equipment, medical products, IT, consumer durables and marketing. Prospective translators should complete a written application form and a short translation test. Experience and educational requirements vary; great emphasis is placed on test results. The company also provides desktop publishing services. Maintains a pool of 300 translators, editors, proofreaders, narrators, and word-counters, all freelance.

OneWorld Language Solutions, 2909 Cole Ave., Ste. 300, Dallas, TX 75204 Phone: (214)871-2909 Fax: (214)871-2907

e-mail: flctans@connect.net Website: www.oneworldlanguage.com
Contact: Gabriele Hayes, Director
Founded 1990. Member: ATA, META. Uses freelance translators and interpreters. Accepts unsolicited resumes. Requires experience, prefers ATA members. Resumes are entered in database and applicants are contacted when the need arises. Translates (both ways) mainly Spanish, French, German, Italian, and Russian. Also Japanese, Dutch, and Chinese. Provides interpretation and language training. Maintains a pool of over 400 translators.

Openworld, 2673 NE University Village, Ste. 1, Seattle, WA 98105
Phone: (206)523-1100 Fax: (206)522-3345
e-mail: orders@openworldtranslations.com
Website: www.penworldtranslations.com
Contact: Helen Tereshina Member: ATA

Origin - The Language Agency, P.O. Box 1648, Makawao, HI 96768
Phone: (808)573-1453 Fax: (808)573-6446
e-mail: info@origin.to Website: www.origin.to
Contact: Chelsea Hill
Founded 1998. Member: ATA, Aquarius, Proz.com. Accepts resumes via website. Translates mainly Chinese, Korean, Indian Languages, Spanish, French, Russian, German, Japanese, Ukrainian and Romanian. Mostly medical, computer-related, business, marketing, telecommunications, tourism, pharmaceuticals and automotive. Has a pool of 600 translators.

OSTrans, LLC 1950 Addison St., Ste. 101, Berkeley CA 94704
Phone: (800)803-9896 Fax: (510)843-5603
e-mail: okada@ostrans.com Web site: www.ostrans.com
Contact: Robert G. Sellin
Founded 1996. Member: ATA, STC. Uses freelance translators and interpreters. Accepts unsolicited resumes. Resumes are reviewed and graded. Work sample are welcome. Requires e-mail. Translates all major Asian and European languages. Focuses on patents and other technical, scientific, pharma/biotech, and legal subjects.

O'Sullivan Menu Corporation, 110 Triangle Blvd., Carlstadt, NJ 07072
Phone: (201)507-1449 Fax: (201)507-4920
Contact: Adriana Marton Member: ATA

-P-

PSC, Inc., 12330 Pinecrest Rd., Reston, VA 20191
Phone:(703)716-5000 Fax:(703)716-5005
e-mail: translat@pscusa.com
Contact: Fred Lothrop
Founded 1988. Member: ATA. Translates Spanish (all dialects) to and from English, and Arabic, Chinese (all dialects), Farsi, Haitian Creole, Russian, Serbo/Croatian, Japanese, and Central and West African languages into English. Emphasis on technical (computer, communications, radar), legal, and business translation. Specialty in transcription and translation for law enforcement agencies. Translators must have a specialized area and security clearance before being matched with an appropriate job. Unsolicited resumes for freelance work are processed through internal channels of selection. PSC, Inc. is a privately-owned firm with a major business area, also providing foreign language services for the private and public sectors. Maintains a pool of 1500 linguists.

Pacific Dreams, Inc., 525 Glen Creek Rd., #230 NW, Salem, OR 97304
Phone: (503)588-7368 Fax: (503)588-7549
e-mail: pacific.dreams@viser.net Web site: www.pacificdreams.org
Contact: Ken Sakai Member: ATA

Pacific Interpreters, Inc., 520 SW Yamhill, #320, Portland, OR 97204
Phone: (800)311-1232 Fax: (800)881-2565
e-mail: recruitment@pacificinterpreters.com

Website: www.pacificinterpreters.com
Contact: Joana Laitinen
Founded 1992. Member: ATA, SOMI, NOTIS, NAJIT, ASTM. Accepts resumes. Translates mainly Spanish, Vietnamese, Chinese, Cambodian, Hmong, Laotian, Arabic and Farsi. Mostly health care. Maintains pool of about 100 translators.

Pacific Ring Services, Inc., 1143 Christina Mill Drive, Newark, DE 19711
Phone: (302)269-1518 Fax: (302)269-1618
e-mail: pacific@dca.net Web site: www.pacificring.com
Contact: Motoko Yuasa, Registered US Patent Agent
Member: ATA. Pacific Ring House, Inc is a patent house.

Pangea Lingua, 3620 N. Washington Blvd., Indianapolis, IN 46250
Phone: (317)920-1600 Fax: (317)920-1601
e-mail: staff@pangealingua.com Website: www.pangealingua.com
Contact: Alessandra Matas
Founded 1994. Member: ATA. Accepts resumes. Translates mainly Spanish, Japanese, German, Chinese, French, Dutch, Portuguese and Russian. Mostly medical, pharmaceutical, promotional, legal and automotive. Maintains pool of over 1000 translators.

Paragon Language Services, 5657 Wilshire Blvd., Ste. 310, Los Angeles, CA 90036 Phone: (323)966-4655 Fax: (323)651-1867
e-mail: info@paragonls.com
Web site: www.paragonls.com
Contact: Hanne R. Mintz Member: ATA, ALC

Peritus Precision Translations, Inc., 727 Industrial Rd., #710, San Carlos, CA 94070 Phone: (650)631-6667 Fax: (650)631-6664
e-mail: info@peritustranslations.com Website:
www.peritustranslations.com
Contact: Dagmar Dolatshko Member: ATA

Phoenix Translations, 2110 A White Horse Trail, Austin, TX 78757
Phone (512) 343-8389 Toll-free (877)452-1348 Fax (512) 343-6721
email: phoenixtranslations@ev1.net www.phoenixtranslations.com
Contact: Jennifer Ruggio

Precision Translating Services, 150 W. Flagler St., Penthouse II, Miami, FL 33130 Phone: (305)373-7874 Fax: (305)381-7874
e-mail: vince@pretran.com Website: www.pretran.com
Contact: Vicente J. de la Vega Member: ATA

Premier Translation Services, Ltd., 62 Stonicker Drive, Lawrenceville, NJ 08648 Phone: (609) 538-9333 Fax:(609)530-0079
e-mail: rzarelli@premiertranslationsusa.com
Web site: www.premiertranslationsusa.com
Contact: Renée Zarelli, Esq., President
Founded 1994. Member: ATA. Accepts resumes. The language pairs we translate are French into English, Italian into English, and Spanish into English. Only translate legal and business documents.. Prospective translators should be ATA members, accredited if applicable. Three+ years full-time experience, college education or higher. Equipment should include WP or Word and modem. Resumes are reviewed on a monthly basis. The company specializes in using lawyers and individuals with a strong legal background, in both U.S. and other legal systems, for translation and editing. Freelance translators do not have to be members of ATA, or accredited by ATA, for consideration. Three or more years' full-time translating experience is desirable. Maintains a pool of 50 translators.

Princeton Technical Translation Center, 333 Bolton Rd., East Windsor, NJ 08520 Phone: (609)443-6770 Fax: (609)443-6778
Contact: Charles Teubner, Director
Founded 1982. Member: ATA, STC. Translates French, German, Spanish, Italian, Portuguese, Chinese, Japanese, Russian and Scandinavian languages, both from and into English. Prospective translators should have advanced technical degrees and be

proficient in Word/WordPerfect and related software.

Professional Translating Services Inc., 44 West Flagler St., Ste. 540, Miami, FL 33130 Phone: (305)371-7887 Fax: (305)381-9824
e-mail: translate@protranslating.com Website: www.protranslating.com
Contact: Luis A. de la Vega
Founded 1973. Member: ATA. Accepts resumes. Translates mainly Spanish, Portuguese, French, Italian, German, Japanese, Russian, Chinese, Arabic and Hebrew. Translations are in legal, technical, financial and advertising. Has a pool of over 350 translators.

ProZ.com, 1448 Madison St., Apt. 410, Oakland, CA 94612
Phone: (510)682-3231 e-mail: henry@proz.com Website: www.proz.com Contact: Henry J. Dotterer Member: ATA

-Q-

Quality Interpreting, 367 University Ave., Ste. 4, Saint Paul, MN 55103 Phone: (651)244-0066 Fax: (651)312-0270
Contact: Paul Yang Member: ATA

Quantum, Inc., 240 S. 9th St., Philadelphia, PA 19107
Phone: (215)627-2251 Fax: (215)627-5570
e-mail: www.quantumtrans@msn.com
Contact: Quan Pham Member: ATA

Quintana Multi-Lingual Services, Inc., 3050 W. Cermak Rd., Chicago, IL 60623 Phone: (773)277-0000 Fax: (773)277-1804
e-mail: quintana@compuserve.com
Contact: Gloria K. Quintana

-R-

RIC International, Inc., 955 Massachusetts Ave., PMB 342, Cambridge, MA 02139 Phone: (617)666-4555 Fax: (617)666-8896
e-mail: r-i-c.com Website: www.r-i-c.com Member: ATA

R.R. Donnelley Financial Translation Services, 75 Park Place, New York, NY 10007 Phone: (212)341-7416 Fax: (212)341-7532
e-mail: FBUTRansServ@rrd.com
Contact: Aouck F. Le Fur
Member: ATA

Ralph McElroy Translation Company, PO Box 4828, Austin, TX 78765 Phone: (512)472-6753 Fax: (512)472-4591
e-mail: tc@mcelroytranslation.com
Website: www.mcelroytranslation.com
Contact: Patricia Bown
Founded 1968. Member: ATA, STC. Accepts resumes. Translates mainly German, Japanese, French, Spanish, Italian, Portuguese, Russian, Chinese, Dutch and Korean. Mostly engineering, technical, medical, legal, pharmaceutical, chemical, patents, localization and high tech. Several hundred active translators.

Rancho Park Publishing, 2203 Balsam Ave., Los Angeles, CA 90064
Phone:(310)470-7488 Fax:(310)470-7930
e-mail: sales@ranchopark.com Website: www.ranchopark.com
Contact: Stan Cheren
Founded 1988. Translates Spanish, French, German, Portuguese and German. Mostly sales and marketing, books and brochures.

Prospective translators should have 10+ years experience and full computer /online capabilities. Resumes are filed for future projects. The company also has desktop publishing capabilities. Maintains a pool of over 700 translators.

Rapport International, 3 Stonegate Lane, #3 Middleboro, MA 02346 Phone: (508)946-3443 Fax: (508)946-3533
e-mail: rapport@rapportintl.com Website: www.rapportintl.com
Contact: Lisa Gavigan Member: ATA
Rennert Bilingual Translations, 216 East 45th St., 17th Fl. New York, NY 10025 Phone:(212)867-8700 Fax:(212)867-7666
e-mail: translations@rennert.com Website: www.rennert.com
Contact: Mikael Poulsen, Director
Founded 1973. Member: ATA, New York Circle of Translators. Translates all languages. Emphasis on all aspects of Patent Law, Marketing, Public Relations, Business and Finance, Videos and Educational Material translation, and website localization. Prospective translators should possess training and experience in the specific fields. Resumes are reviewed for education and experience, and a Rennert application is sent to the translator for completion. Maintains a pool of 3000 translators.

Rescribe, P.O. Box 503981, San Diego, CA 92150
Phone: (858)487-5292 Fax: (858)487-5292
e-mail: erin@rescribe.com Website: www.rescribe.com
Contact: Erin Berzins
Founded 2001. Member: ATA. Accepts resumes. Translates mainly Spanish, French, Chinese, Vietnamese, Japanese, Italian, German, Korean and Russian. Mostly legal, insurance, medical, finance and insurance. Has a pool of several thousand translators.

Richard Schneider Enterprises, Inc., 27875 Berwick Drive, Suite A, Carmel, California 93923
Telephone: (800)500-5808 x25 Fax: (831)622-0524
e-mail: san-dra@idioms.com Website: www.idioms.com
Contact: Richard A. Schneider Member: ATA

Rosetta, 2261 Market St. #318, San Francisco, CA 94114
Phone: (415) 550-9151 Fax: (415)550-9162 e-mail: oshea@sirius.com
Contact: Christina O'Shea, Director
Founded 1995. Member: ATA. Translates Japanese, Chinese, French, German and Korean, both from and into English, and Italian, Spanish, Thai, Swedish, Norwegian, Polish, Danish, Dutch, Portuguese and Russian into English. Emphasis on Computer (hardware and software), Semiconductors, Chemical, Medical, Pharmaceutical, Automotive, Patent, Business, and Marketing translation. Prospective translators should have at least five years experience as translators, plus five years experience in a technical field or in business, and a degree in their area of specialization. Resumes are entered into a data base and sent a test if they fulfill company requirements. The company hires translators with technical specialties as opposed to "generalists" and has every document reviewed for accuracy by a technical editor. Maintains an "active" pool of 50 translators.

Rosetta Stone Associates, Inc., 34 Franklin St., Ste. 200A, Nashua, NH 03063 Phone: (603)883-9388 Fax: (603)595-8673
e-mail: translate@rosettastoneinc.net Website: www.rosettastoneinc.net
Contact: Jana Coughlin
Founded 1972. Member: ATA, STC. Accepts resumes. Translates mainly French, Italian, German, Spanish, Russian, Chinese, Japanese, Arabic, Portuguese and Greek. Mostly industrial, medical, legal, scientific, marketing, computers, software and localization. Uses pool of over 2000 translators.

Russian and Slavic Language Services, 271 Madison Ave., 3rd Fl., New York, NY 10016 Phone: (212)481-4980
Fax: (212)683-4801
e-mail: karina@russiantranslation.com
Website: www.russiantranslation.com
Contact: Karina Gukasyan Member: ATA

RussTech, 1338 Vickers Rd., Tallahassee, FL 32303

Phone: (850)562-9811 Fax: (850)562-9815
e-mail: Contact: Cynthia Seaborn Member: ATA

-S-

S.B.F. TRANSLATIONS INC., 5601 Collins Ave #1021, Miami Beach, Florida 33140 Phone: (305) 866-6079
Fax: (305) 866-6162
e-mail: SBFTRANS@AOL.COM
Contact: Silvia B. Fernandez
Founded 1994. Member: ATA. Accepts resumes. Translates Spanish, Portuguese and Creole. Mostly legal, medical, immigration and Christian books. Has three translators.

SH3, Inc., 5338 East 115th St., Kansas City, MO 64137
Phone: (816)767-1117 Fax:(816)767-1727
e-mail: chubbard@sh3.com Website: www.sh3.c-om
Contact: Cathy Hubbard, General Manager
Founded 1980. Member: ATA, STC. Translates mainly French, German, Spanish, Italian, Dutch, Portuguese, Swedish, Chinese and Japanese. Emphasis on Agricultural and Industrial Equipment, and Consumer Equipment translation. Prospective translators should have Windows-based computer, modem and fax; they should have experience in the agriculture, industrial and mechanical areas. Translator's Workbench is a definite plus. Resumes are reviewed by subject specialty, experience, and language, and flagged for a future contact. Does around 20 million words a year. The company deals in high volume, highly technical manuals. Maintains an active pool of 70 translators.

Sajan, Inc; 215 N 2nd St, River Falls, WI 54022 Phone: (715)426-9505 Fax: (715)426-0105 e-mail: info@sajan.com
Web Page: www.sajan.com; Contact: Angel Zimmerman, Director of Operations
Member: ATA; Founded in 1997. They translate English in to European Spanish, Latin American, European Portuguese, European French, Canadian French, German, Italian, Norwegian and Swedish. They also Translate English into all Western European Languages, all Eastern European Languages and all Asian Languages. Main subjects are medical, medical devices, engineering, technical, dentistry, and general. They specialize in translation services and editing/proofing. Accepts freelance resumes. They require 5 or more years translating experience and translating in particular fields. Also ATA accreditation or equivalent in target language country. Their annual work_load is about 2,500,00 words per year. They have 300-400 translators in their pool.

SARJAM Communications, Ltd., 4370 NE Hulsey St., Ste. 122, Portland, OR 97213 Phone:(503)287-9277
Fax:(503)287-9277 e-mail: 70314. 1226@compuserve.c-om
Contact: Seth A. Reames
Founded 1987.Member: ATA. Translates Japanese, Korean, Chinese, Spanish, German, French, Italian and Russian, both from and into English. Translates all subjects. Prospective translators should have 5+ years experience, a computer (MAC or PC), fax and modem. Resumes are examined and filed by language. The company also offers desktop publishing and graphic design. Their in-house computers are Macintosh. Maintains a pool of 40-50 translators.

Schreiber Translations, Inc., 51 Monroe St., Ste. 101, Rockville, MD 20850 Phone:(301)424-7737 Fax:(301)424-2336
e-mail: translation@schreibernet.com Website: www.schreibernet.com Contact: Karen Ffinch, Director, Translation Services
Founded 1984. Member: ATA. Uses freelance translators and interpreters. Accepts unsolicited resumes only on website. Translates (both ways) mainly Japanese, Russian, German, French, Spanish, Chinese, Italian, Arabic, Korean, Vietnamese, Hebrew, Polish, and Dutch, and close to 90 additional languages and dialects. Emphasis on patents, aerospace, law, medicine, communications, computers, engineering, public relations, military and maritime subjects, chemical subjects, and business and finance. Maintains a pool of over 1300 translators - 350 active, and 200 of those very active. In addition to translation and interpretation, the company produces multilingual brochures and advertisements, prints manuals and business literature in a variety of multinational fonts, and does voiceovers for video and films. Maintains a pool of over 1300 translators—350 active, and 200 of those very active.

SinoMetrics International, Inc., 3503 188th St. SW, Lynnwood, WA 98115 Phone: (425)776-5648 Fax: (425)776-6648
email: info@sinomet.com Website: www.sinomet.com

Member: ATA

Speak Easy Languages, 757 South Main St., Plymouth, MI 48170 Phone:(734)459-5556 Fax:(734)459-1460
e-mail: selanguages@earthlink. net
Contact: Cristina Clark
Founded 1980. Member: ATA. Translates all major language pairs. Emphasis on Automotive business, Brochure, Advertising and Legal translation. Prospective translators should be ATA accredited and possess fax and modem capabilities. Resumes are filed and the applicant called upon as the circumstance arises. The company also provides talent for voiceover work. Maintains a pool of 100.

State Farm Insurance, One State Farm Plaza, SC-2, Bloomington, IL 61710 Phone: (309)766-3760
Contact: Joan Epperson Member: ATA

Superior Translations, 1924 Minnesota Ave., Duluth, MN 55802
Phone: (218) 727-2572 Fax: (218)727-2572
e-mail: towardinc@aol.com Web site: www.isuperiortranslations.com
Contact: Elisa A. Troiani
Founded 1993. Member: ATA. Uses freelance translators and interpreters. Accept unsolicited resumes. Resumes for translation and interpretation should be mailed. Applicants are carefully screened and required to submit samples, references, and tests. Requirements are honesty, accuracy, editing/proofing skills, reasonable rates, punctuality, and an eye for detail. Translates mainly Spanish, French, German, Japanese, Chinese, Russian, Portuguese, Italian, Polish and Hmong. Emphasis on all subject areas. Has 250_300 translators in its pool.

Suzuki, Myers & Associates, Ltd., P.O. Box 852, Novi, MI 48376 Phone:(248)344-0909 Fax:(248)344-0092
e-mail: language@suzukimyers.com Website: www.suzukimyers.com
Contact: Kumiko Oh
Founded 1984. Member: ATA, MITN, Japan-America Society. Translates Japanese and Korean only, both from and into English. Emphasis on Manufacturing-related translation (Autos, Engineering, Quality Design, Robotics, Plastics, Assembly, Electronics, Paint, Contracts, Product Liability).Prospective translators should be competent and diligent. Resumes are responded to with a phone call. The company provides translation, interpretation, and consultation services in North American-Japanese relations. They also provide DTP services. Maintains a pool of about 40 translators.

-T-

Techno-Graphics and Translations, Inc., 1451 East 168th St., South Holland, IL 60473 Phone: (708)331-3333 Fax: (708)331-0003
e-mail: techno@wetrans4u.com Website: www.wetrans4u.com
Contact: Pinay Gaffney
Founded 1972. Member: ATA. Uses freelance translators and interpreters. Accepts unsolicited resumes. Resumes are entered in database. Requires native ability and technical background. Translates (both ways) mainly European, Asian, and Middle Eastern languages. Main areas are telecommunications, agriculture, industry, automotive, and medical. Worldwide pool of 200 translators.

TechTrans International, Inc., 2200 Space Park Drive, Ste. 410, Houston, TX 77058 Phone: (281)335-8000 Fax: (281)333-3503
e-mail: techtrans@tti-corp.com Website: www.tti-corp.com
Contact: Beth Williams Member: ATA

Techworld language Services, Inc., 1250 W. 14 Mile Rd., Ste. 102, Clawson, MI 48017 Phone: (248)288-5900 Fax: (248)288-7900
e-mail: info@techworldinc.com Website: www.techworldinc.com
Contact: Fred Meinberg Member: ATA

Telelingua International; 46 Sayre Drive, Princeton, NJ 08540
Phone: (609)951-9511 Fax: (609)951-0550
e-mail: lmellet@telelingua.c-om Web Site: www.telelingua.c-om

Contact: hbaraona@telelingua.com
Member: LISA, GALA; Telelingua International was founded in 1985. Jean-Didier Boucau is the CEO of the company. They translate English to FIGS, Japanese, Korean, Chinese, Dutch, French and Spanish. The main subjects that they translate are ERP software, medical/pharma, banking and manufacturing. They do not use freelance translators. They do accept resumes. The resumes are human resources processed. You must be certified. Telelingua is headquartered in Brussels, Belgium and has sales and manufacturing operations in Paris, France and Princeton, NJ. It also has strategic alliances in Latin America and Asia, thus providing it's customers with a global solution to its GILT requirements. Telelingua is also recognized for its leadership in developing cutting edge translation technologies. T-Remote Memory, its best-selling application, is a web-services based Translation Memory technology allowing multiple remote translators to work on the same job at the same time using the same TMs. They have approx. 1,500 translators in their pool.

TermSeek Inc./Lingotek; 1257 W. 75 N., Centerville, Utah 84014
Phone: (801)294-7700 Fax: (801)294-7711
e-mail: hr@termseekinc.com Web Page: www.termseekinc.c-om
Contact: Charles
Member: ATA; TermSeek Inc./Lingot-ek was founded in January of 2001. They translate English to Spanish, Chinese, Japanese, German, French, Italian, Bosnian, Russian, Arabic and Punjabi. They also did work in over 50 language pairs.The main subjects are medical, legal, engineering, manuals, manufacturing, marketing and genomics. They use freelancers. They also accept resumes. Applicants need to list native language and their rates. They typically employ 4 independent translators on every job. They also develop software to assist their translators in doing the work at a higher level of quality and speed. They currently have about 3,000 translators in their pool.

TerraSpan, Inc., 12647 Alcosta Blvd., Ste. 150, San Ramon, CA 94583
Phone: (925)866-3030 Fax: (925)866-3036
e-mail: galiloupour@terraspan.com Website: www.erraspan.com
Contact: Gloria Aliloupour Member: ATA

TEXTnology CORPoration, 20 Trafalgar Square, Ste. 456, Nashua, NH 03063 Phone:(603)883-8398 Fax:(603)883-8998
e-mail: nadeau@textcorp.com Website: www.textcorp.com
Contact: Ronald Nadeau Member: ATA

Thomson Scientific, 3501 Market Street, Philadelphia, PA 19104
e_mail: ellen.boyar@thomson.com Web site: www.thomsonisi.com
Contact: Ellen A. Boyar

3 Way Talk, 2255 Skyfarm Drive, Hillsborough, CA 94010
Website: www.3waytalk.com
Contact: Dr. B. Goldstein Member: ATA

3DWord, Inc., 4325 Athens High Way, Ste. 4-13, Loganville, GA 30052
Phone: (770)613-0879 Fax: (770)849-9627
e-mail: julieMacgregr@3word.com Website: www.3word.com
Contact: Julie MacGregor Member: ATA

TLS - Translation Service, 69 Broad St., Red Bank, NJ 07701
Phone: (732)530-4542 Fax: (732)849-6755
e-mail: trans90@aol.com
Contact: Ingeborg Perndorfer
Founded 1982. Member: ATA. Uses freelancers and accepts unsolicited resumes. Resumes are checked and filed until needed. Applicants must show experience. Translates all languages and all subjects. Maintains a pool of 30-50 translators.

Total Benefit Communications, Inc., 5775-B Glenridge Drive, Ste. 120, Atlanta, GA 30328 Phone: (404)256-5042
Fax: (404)256-5101
Contact: Anne McKillips Member: ATA

Total System Services, Inc., 1600 First Ave., TSYS Corporate Campus, 1st Fl., #1318, Columbus, GA 31901 Phone: (706)649-5450 Fax: (706)649-8148 Contactl: Evan Latham.

Toward, 1924 Minnesota Ave., Duluth, MN 55802
Phone: (218)727-2572 Fax: (218)727-2572
e-mail: towardinc@aol.com Web site: www.superiortranslations.com
Contact: Dr. Elisa A. Troiani Member: ATA

TRADOS Corporation, 113 S. Columbus St., Ste. 400, Alexandria, VA 22314 Phone: (703)683-6900 Fax: (703)683-9457
e-mail: edith@trados.com Website: www.trados.com
Contact: Edith Westfall Member: ATA

Traducciones LinguaCorp, Poba International, #1-20072, P.O. Box 02-5255, Miami, FL 33102 Phone: 58(2)7625745
Fax: 58(2)7625745
e-mail: linguacorp@compuserve.com
Contact: Francisco J. Pance Member: ATA

Trans Global Translation & Immigration Services, 175 Fontainbleau Blvd., Ste. 2G 8, Miami, FL 33172
Phone: (305)552-9793 Fax: (305)223-4080
e-mail: trnsglobal@aol.com Member: ATA

TransACT Inc., 8423 Mukilteo Speedway, Ste. 100, Mukilteo, WA 98275 Phone:(206)348-5000 Fax:(206)348-3095
e-mail: support @transact.com Website: www.transact.com
Contact: Ms. Tomi Renshaw
Founded 1994. Member: ATA, NOTIS, Washington Software Assn., ASCD. Translates Spanish, Russian, Cambodian, Korean, Chinese, Lao, Thai, Arabic, Japanese and Vietnamese from English. Emphasis on Education, Medical and Utilities translation. Prospective translators must have computer skills and PageMaker experience. Resumes are reviewed and applicants contacted if needed. The company specializes in translation and typesetting for school districts in the US. Their software division creates custom CD-ROMs for Mac/Win in all languages. Maintains a pool of 20 translators.

Trans-Caribe Communications, Inc., 9109 Queen Elizabeth Court, Orlando, FL 32818 Phone: (407)260- Fax: (407)260-6349
e-mail: transcc@aol.com Website: www.iotrak.com/trans-caribe.html
Contact: Sherry E. Allen-Diaz Member: ATA

Transemantics, Inc., 4301 Connecticut Ave. NW, Ste. 146, Washington, DC 20008 Phone: (202)686-5600 Fax: (202)686-5603
Contact: M-L Wax Cooperman Member: ATA

Transimpex Translations, 8301 East 166th St., Belton, MO 64012 Phone: (816)561-3777 Fax:(816)561-5515
e-mail: transmpx@sound.net Web site: www.transimpex.com
Contact: Ingrid Pelger, Doris Ganser or Brian White
Founded 1974. Member: ATA, Board member International. Trade Club of Greater Kansas City. Uses freelance translators and interpreters. Accepts unsolicited resumes. Resumes are put in database until needed. Requires degree in linguistics or translation. Main languages translated are German, French, Spanish, Portuguese, Italian, Japanese, Russian, and Arabic. Translates all subjects, with long experience in localization. Has overall pool of some 5000 translators, with about 150 active ones.

Transfirex Translation Services, Inc., 12509 Village Sq. Terr., #201, Rockville, MD 20852 Phone: (301)984-9753
Fax: (630)604-8323
e-mail: info@transfirex.com Website: www.transfirex.com

Contact: Margaret Johnson
Founded 2001. Member: ATA. Accepts resumes. Main languages are Turkish, German, Spanish, Italian, French and Portuguese.Mostly scientific, engineering, social sciences, and safety, emergency and rescue. Has a pool of 12 translators.

TRANSGLOBAL Language Services, 534 Princeton Cove, Memphis, TN 38117 Phone: (901)680-0927
Contact: Robyn Hassell

Translate4me Inc.; 9720 Wilshire Boulevard, Suite 205, Beverly Hills, CA 90212; Phone: (310)274-9771 Fax: (310)274-9431
e-mail: translation@translate4me.com Web Page: www.translate4me.com Contact: Olessia Laird
Member: ATA, ITI, BBB. The main languages that they translate are English to French, German, Spanish and Italian. They also translate into 80 languages. They translate into all subjects. Translate4me Inc does use freelance and they do accept resumes. You may send your resume to olessia@translate4me.com. Suitable applicants are sent an application form, terms of business nad are reference checked. They require that applicants have graduated and have at least 5 years translation experience. They should also have professional translator's qualifications. They have about 3,000 translators in their pool.

Translation Aces, Inc., 29 Broadway, Ste. 2301, New York, NY 10006 Phone: (212)269-4660 Fax: (212)269-4662
Contact: Serge Nedeltscheff
Founded 1952. Member: NY Circle, NY Chamber of Commerce. Translates Spanish, French, German, Japanese Italian, Portuguese, Russian and Greek, both from and into English, as well as *all* other language pairs. Emphasis on Legal, Patent, Medical Financial, Scientific, Advertising, Engineering, and almost all other areas of translation. Prospective translators should have "experience, experience, experience." Resumes are responded to with the company's application form. The company has 45 years experience in providing the legal, financial, advertising, medical and governmental communities with translation, interpreting, voiceover and consulting services. Maintains a pool of about 400 translators.

Translation Company of America, Inc., 10 West 37th St., New York, NY 10018 Phone:(212)563-7054 Fax:(212)695-2385
e-mail: trude@tcany.com Website: www.languagematters.com
Contact: Trudi Mathys Member: ATA

Translation Company of New York, 8 South Maple Ave., Marlton, NY 08053 Phone: (856)983-4733 Fax: (856)983-4595
e-mail: tcny2000@cs.com Web site: www.tcny.net
Contact: Jennifer C. Thompson

Founded 1983. Member: ATA. Uses freelance translators and accepts unsolicited resumes. Translates mainly French, German,Spanish, Italian, Japanese, Portuguese, Dutch, Russian, Chinese and Swedish. Main areas are pharmaceutical, chemical, medical, engineering, and electronics. Does about 7_8 million words a year. Has active pool of 200 translators.

Translation Services, Inc., 5860 Forward Ave., Office #1, Pittsburgh, PA 15217 Phone: (412)242-8240 Fax: (412)242-1241
e-mail: ATA-info@translations.us.com
Website: www.translations.us.com
Contact: Lisa M. Heinrich Member: ATA

Translation Services International, 1100 South Mint Street, Suite 107, Charlotte, NC 28203 Ph: (704)347-5882
Fax: (704)347-5899
e-mail: info@tsitranslation.com Web: www.tsitranslation.com
Contact: Victor M. Guzman, President/CEO

Translations & Language Consulting, 8105 Hearthstone Place, Antelope, CA 95843 Phone: (916)729-4222 Fax: (916)729-4232
e-mail: emorrow@trans-language.com
Website: www.trans-language.com

Contact: Elena Morrow   Member: ATA

Translations Unlimited, 1455 Forest Hill SE, Grand Rapids, MI 49546 Phone:(616)942-5742 Fax: (616)526-8583
e-mail: lmathews@calvin.edu Contact: Leslie Mathews, Director
Founded 1980. Member: ATA. Accepts resumes. Main languages are French, Spanish, German, Japanese, Portuguese, Italian, and Vietnamese. Main areas are patents, sales literature, video scripts, and voice-overs, employee handbooks, sales literature, legal documents, and technical specifications. Translator pool of 25-30.

translations.com, 3 Park Ave., 37th Fl., New York, NY 10016
Phone: (212)689-1616 Fax: (212)685-9797
e-mail: LRC@translations.com Website: www.translations.com
Contact: Pamela Kraljevich
Founded 1999. Member: ATA. Accepts resumes. Translates mainly Spanish, French, German, Chinese, Japanese, Portuguese, Italian, Korean, Norwegian and Swedish. Mostly software, IT-Telecom, finance, travel and medical. Provides full suite of globalization services. Pool of 500 translators.

Translation3, 49 Union Park, #2, Boston, MA 02118
Phone: (617)482-2223 Fax: (617)482-2121
e-mail: jjtranslation3@translation3.com Website: www.translation3.com
Contact: Jacqueline M. Jacquiot Member: ATA

The Translators, PO Box 303, Topsfield, MA 01983
Phone: (978)887-9234 Fax: (978)887-6657
e-mail: trnslate@shore.net Web site: www.thetranslatorsinc.com
Contact: Jim Swift, Personnel Manager
Founded 1980. Accepts resumes. Translates all languages, mostly medical, patents, human resources, computer, sci-tech, legal, electronic and engineering. Unlimited translator pool.

Translingua, Inc., 5457 Twin Knolls Rd., Columbia, MD 21045
Phone: (410)730-9700 Fax: (410)730-9736
e-mail: moneill@translingua.com Website: www.translingua.com
Contact: Mary J. O'Neill
Founded 1984. Member: ATA, NCATA, STC, CATI. Translates French, Italian, German, Spanish, Japanese, Chinese, Korean, Swedish, Danish, Portuguese, Russian, Hebrew, and Arabic from English. Emphasis on technical, scientific, medical, financial, software, packaging and labeling, marketing and advertising. Specialty in full-service graphics and high resolution double byte output. Prospective translators should have an area of expertise as well as a computer, email, modem, and fax. Resumes are screened, evaluated and databased. The company's translations, often large and multilingual, are aimed at publication and the end user. It provides services for the entire process from project management to printing. Maintains a pool of more than 1000 translators.

Transperfect Translations International, Inc., 3 Park Ave., 28th Fl., New York, NY 10016 Phone: (212) 689-5555
Fax: (212)251-0981
e-mail: info@transperfect.com Website: www.transperfect.com
Contact: Information Member: ATA

Transtek Associates, Inc., 599 North Avenue, Door 9, Wakefield, MA 01888 Phone: (781) 245-7980 Fax: (781)245-7993
e-mail: michele@transtekusa.com Website: www.transtekusa.com
Contact: Michele Phillips
Founded 1964. Member: ATA. Uses freelance translators and interpreters. Accepts unsolicited resumes.

Tristan Translations, 1130 Taylor St., San Francisco, CA 94108 Phone:(415)474-2611
Fax:(415)928-4755 e-mail: tristant@ix.netcom .com
Contact: Marc de Tristan, Jr., Owner.
Founded 1978. Accepts resumes. Translates Spanish, German, Chinese, Japanese, Portuguese, French, Italian, and numerous

other languages, both from and into English. Emphasis on Patents, Technical Document, Proposals and RFQ, Discovery Document and Insurance Claim translation. Prospective translators should exhibit fluency in the English resume or phone contact, education and word processing and file transfer skills. Resumes are read with an eye to flaws, and if approved, entered into a database for future work. Maintains a pool of over 100 translators.

Tru Lingua Language Systems, Inc., 2081 Business Center Drive, Irvine, CA 92626 Phone: (949)955-1151 Fax: (949)955-1153
e-mail: Trulingua@aol.com
Contact: Diana Knapstein
Founded 1984. Accepts resumes. Translates mainly Spanish, French, German, Dutch, Portuguese and Japanese. Mostly technical, medical and lega. Prospective translators should be ATA certified, ATA members, and possess wide experience. Maintains a pool of over 350 translators.

Trustforte Language Services, 271 Madison Ave., 3rd Fl., New York, NY 10016 Phone: (212)481-4980 Fax: (212)481-4972
e-mail: aschroeder@trustforte.com
Website: www.trustfortelanguages.com
Contact: Jackie Cordero
Founded 1992. Member: ATA. Accepts resumes. Works in most languages. Provides translators and interpreters to a variety of fields, from business to entertainment.

turkishtranslation.com, MSR Consulting Ltd., 33 N. High St., Clinton, CT 06143 Phone: (860)669-2777 Fax: (909)363-9150
e-mail: info@turkishtranslation.com Website: www. turkishtranslation.com
Contact: Asiye Yucel Member: ATA

-U-

Uniscape, Inc., 1292 Hammerwood Ave., Sunnyvale, CA 94089
Phone: (408)743-3500 Fax: (408)743-3600
Website: www.uniscape.com
Contact: Seda P. Mansour Member: ATA

United International Services, Inc., 12701 W. Warren, Dearborn, MI 48126 Phone: (313)846-6903 Fax: (313)846-4278
Contact: Faye Awada Member: ATA

United Nations Translators & Interpreters, Inc., 1515 E. Livingston St., Orlando, FL 32803 Phone: (407)894-6020 Fax: (407)894-6693
e-mail: unti@unti.com Website: www.unti.com
Contact: Fiona Como
Founded 1991. Member: ATA. Accepts resumes. Translates mainly Spanish, Creole, French, German, Arabic, Farsi, Italian, Portuguese and Vietnamese. Mostly legal, medical, technical and INS documents. Uses about 50 translators.

Universal Dialog, 4150 Murphy Canyon Road, Suite 210, San Diego, CA, 92109 Phone: (858) 273 9060 Fax: (858) 273 9062
e-mail: email@universaldialog.com Web site: www.universaldialog.com
Contact: Alexandra Briskin Member: ATA

University Language Center, Inc., 1313 Fifth St. SE, Ste. 201, Minneapolis, MN 55414 Phone:(612)379-3574 Fax:(612)379-3832
e-mail: translation@ulanguage.com Website: ulanguage.com Contact: Therese Shafranski, Translation Department Manager
Founded 1986. Member: ATA. Accepts resumes. Translates Spanish, French, German, Italian, Portuguese, Russian, Chinese, Japanese and Hmong, as well as Turkish, Indonesian, Arabic and all Southeast Asian languages. Emphasis in all technical areas.

The company handles all phases of production from translation to camera-ready art, including cultural assessment of the source documents. Maintains a pool of 250 translators.

University Translators Services LLC, 220 N. 5th Ave, Ann Arbor, MI 48104 Phone: (734)655-7295 Fax: (734)655-1345
e-mail: lfinch@univtrans.com Web Page: www.univtrans.c-om
Contact: Laurie Finch, President
Member: ATA. Founded in 1988. They do accept freelance and unsolicited resumes. The resumes are entered into their database under language, specialty and location. They require that you have 10+ years of experience, certification and/or ATA and/or Master's Degree in Translation/Interpretation. Their main language pairs are English to French, German, Spanish, Portuguese, Chinese, Japanese, Korean, Thai, Arabic, Urdu and Vietnamese. Some of tier subjects are manuals for automotive & manufacturing of all types, patents, immigration related documentation, consumer surveys, promotional brochures and video for sales, court interpreting, subtitling and voice over, website HTML pages and software localization. Their work load is about 50% written translation and 50% interpretation. They have 25 full time translators or 50 full and part time translators. University Translators Services is a woman owned agency operating in the capacity of an LLC.

U.S. Translation Company, 3701 Harrison Blvd., Ogden UT 84403
Phone: (801)393-5300 Fax: (801) 393-5500
e-mail: info@ustranslation.com Website: www.ustranslation.com
Contact: David Utrilla Member: ATA

-V-

Venturi Technology Partners, 400-1 Totten Pond Rd., Waltham, MA 02451 Phone: (781)890-7007 Fax: (781)890-4433
e-mail: mklinger@ventwipartners.com
Web site: www.venturiglobalization.com
Contact: Michael L. Klinger
Founded: 1993. Member: ATA, NETA, STC. Accepts resumes. Translates mainly Japanese, French, Italian, German, Spanish, Korean, Chinese, Portuguese, Hebrew and Arabic. Mostly software, medical and pharmaceutical. Has a pool of 500 translators.

Vital International Programs, Inc., 34514 Dequindre, Ste. C, Sterling Heights, MI 48310 Phone: (586) 795-2500 Fax: (586) 795-5763
e-mail: lnespolo@protranslating.com website: www.vitalinternational.com

Voices for Health, Inc., 894 Fuller NE, #2, Grand Rapids, MI 49503
Phone: (616)458-8388 Fax: (616)458- 8322
e-mail: info@voicesforhealth.com Web site: www.voicesforhealth.com
Contact: Scott Van Til, Translations; Van Nguyen, Interpretation Coordinator.

Founded 1997. Member: ATA. Accepts resumes. Translates mainly Spanish, Vietnamese, Chinese and Arabic. Mostly healthcare and social services. Uses hundreds of linguists.

Vormbrock Translations, Inc., 10415 Jockey Club Drive, Houston, TX 77065 Phone: (281)894-9722 Fax: (281)890-0062
e-mail: VTITrans@pdq.net Website: www.vormbrock.com
Member: ATA

-W-

WKI International Communications, 27 W. 20th St., Ste. 402, New York, NY 10011 Phone: (212)255-6100 Fax: (212)255-8461
e-mail: smwoodruff@wkiintl.com Website: www.wkiintl.com
Contact: Susan Woodruff Member: ATA

Welocalize.com, 241 E. 4th St., Ste. 207, Frederick, MD 21701
Phone: (301)668-0330 Fax: (301)668-0335
e-mail: info@welocalize.com Website: www.welocalize.com
Contact: E. Smith Yewell Member: ATA

Wizards Of The Coast, Inc., P.O. Box 707, Renton, WA 98057
Fax: (425)204-5806 Website: www.wizards.com
Contact: David Serra Member: ATA
Word for Word, Inc., 325 W. Chickasaw Rd., Virginia Beach, VA 23462 Phone: (757)557-0131 Fax: (757)557-0186
e-mail: wrd4wrd@exis.net Contact: Curtis Hovey Member: ATA

Word Magic Software, 12220 Beechnut PMB 2026, Houston, TX, 77072 Phone: (281)564_3022 Fax: (281)568_2143
e_mail: info@wordmagicsoft.com Web Page: www.wordmagicsoft.com Contact: Kristha Serrano
Founded in 1989; Word Magic Software only translates English into Spanish. They translate any subject. They do not use freelance and they do not accept resumes. They have English-Spanish Translation Software, Dictionary, Thesaurus, Spell Checker, Verb Conjugator. They also offer translation services in both languages, English and Spanish.

WORDNET, Inc., 282 Central St., Ste. 2, P.O. Box 976, Acton, MA 01720 Phone: (978)264-0600 Fax:(978)263-3839
e-mail: wordnet@wordnet.c-om Website: www.wordnet.com
Contact: lee Chadeayne
Founded 1985. Member: ATA, German-American Business Council. Uses freelance translators and interpreters. Accepts unsolicited resumes. Resumes are reviews and applicants are sent out forms to fill out, and are periodically tested and evaluated. Main languages translated are French, Spanish, German, Italian, Chinese, Japanese, and Korean. Main areas are technical, medical, machinery, legal, and business/marketing. The company specializes in translation, typesetting and printing, preparing brochures, catalogs, manuals, legal agreements and Web pages.They help companies get started in the international marketplace. Maintains a pool of some 2000 translators.

Worldwide Translations, Inc., 2317 Fossil Canyon, Henderson, NV 89052 Phone: (800)293-0412 Fax:(800)536-6995
e-mail: wwtranslate@cs.com Web site: www.wwtranslations.com
Contact: Robert Sansing
Founded 1994. Member: ATA, RAPS, FDLI, DIA, NAJIT. Accepts resumes. Translates mainly French, Italian, German, Spanish, Dutch, Portuguese, Greek, Swedish, Japanese and Chinese. Emphasis on Medical, Machinery Manufacture and Computer Products translation. Prospective translators should be native_speaking ATA members with 5+ years technical translation experience and references. Resumes are responded to by telephone when work becomes available. The company is expert in the CE Mark and Medical Device areas and specializes in the translation of instructional manuals. They hold their translators to a high standard of accuracy. Maintains a pool of over 100 translators.

The Write Stuff, 11000 Lake City Way, NE, Seattle, WA 98125
Phone: (206)548-1111 Fax: (206)548-9116
e-mail: translations@writestuff.com Website: www.writestuff.com
Contact: Susan Hedding Member: ATA

Wudang Research Association, 2301 South Congress Ave., #1021 Boynton Beach, FL 33426 Phone: (561) 733-1417
e-mail: tmorgan@wudang.com Web site: www.wudang.com
Contact: Theresa M. Morgan

-Z-

Zanzitek, LLC, 475 Commonwealth Ave., Ste. 101, Boston, MA 02215
Phone: (617)236-7791 Fax: (617)236-7792
e-mail: antonio_gangemi@zanzitek.com

FOREIGN TRANSLATION COMPANIES

Argentina

Translationary, Viamonte 524, 2° Ofc. 17, Buenos Aires, 1053 Argentina Phone: 54(11)43130088 Fax: 54 (11) 43115301
e-mail:info@translationary.co-m
Contact: Marcos von Haeften

Australia

Pacific Rim Translation International Pty., Ltd., Level 10, 55 Swanton St., Melbourne, Victoria 3000, Australia
Phone: 61(3)96533012 Fax: 61(3)96548644
e-mail: apak@prti.com.au Website: www.prti.com.au
Contact: Audrey Pak

Belgium

Anthea Languages S.A., 8 Quai Godefroid Kurth, Liege, B4020 Belgium
Phone: 32(4)3444566 Fax: 32(4)3444335
e-mail: info@anthealanguages.com
Website:www.anthealanguages.c-om
Contact: Yves G. Vanneste

Brazil

Amerikas Traducôes. Ltda., Av. Franklin Roosevelt 39, Ste. 1108, Rio de Janeiro, RJ, 20021-120 Brazil
Phone: 55(21)533-5821 Fax: 55(21)533-6853
e-mail: amerikas@tradutores.com
Contact: Leonardo R. Moraes

Canada

All Languages Ltd., 421 Bloor St. E, Ste. 306, Toronto, Ontario
M4W 2T1 Canada Phone: (416)975-5000 Fax: (416)975-0505
e-mail: info@alllanguages.com Website: www.alllanguages.com
Contact: Paul M. Penzo

The Bank of Nova Scotia, 44 King St. W, Toronto, Ontario M5H 1H1 Canada Phone: (416)866-3354 e-mail: joanne.akai@scotiabank.com
Contact: Joanne Akai

Global Translations & Interpreters Services, Inc., 277 Wellington St. W, Ste. 710, Toronto, Ontario M5V 3E4, Canada
Phone: (416)593-1444 Fax: (416)593-6344
e-mail: mailbox@globaltranslations.com
Website: www.globaltranslations.com
Contact: Tim Koressis

HSN Linguistic Services Ltd., 5360 Canotek Rd., Unit 20, Gloucester, Ontario K1J 9ES, Canada
Phone: (613)748-1915 Fax: (613)748-5596
e-mail: dcyr@hsntraslation.com Website: www.hsntraslation.com
Contact: Diane Cyr

in FRENCH only inc./in SPANISH too! Translations, 25 Rockcastle Drive, Toronto, Ontario M9R 2V2, Canada
Phone: (416)248-5648 Fax: (416)614-3806
e-mail: translations@idirect.com Website: www.translations.ca
Contact: Kim Pines

Multicorpora R&D Inc., 490 Saint Joseph Blvd., Ste. 102, Hull, Quebec, J8Y 1G3, Canada
Phone: (819)778-7070 Fax: (819)778-0801
e-mail: info@multicorpora.com Website: www.multicorpora.com
Contact: Gerry Gervais

Parenty Reitmeier Translation Services, 123-B Marion St., Winnipeg,
Manitoba R2H 0T3, Canada
Phone: (204)237-3737 Fax: (204)237-9997
e-mail: Joanne@parenty.mb.ca Web site: www.parentyreitmeier.com
Contact: Joanne Rasmussen
Founded 1992. Member: ATA, STC, SAE, LISA. Accepts resumes. Requires translation into mother tongue and knowledge of Trados. Main languages are French, Spanish, Portuguese, German, Italian, Dutch, Danish, Norwegian, Swedish and Finnish.Mostly mechanical, medical and legal. Maintains pool of 90 translators.

Shoreline Translations, Inc., 100 Waverley St., #103, Ottawa, Ontario K2P 0V2, Canada Phone: (613)237-3503
Fax: (613)237-3503
Contact: Gordon R. McLeod

Transcript Communications, Inc., 625 Ave. du President Kennedy, Bureau 800, Montreal, Quebec H3A 1K2, Canada
Phone: (514)874-9134 Fax: (514)874-9192
e-mail: info@comtranscript.com Website: www.comtranscript.com
Contact: Wilma Scappaticci

China

Beijing E-C Translation & Localization Co., Ltd., P. O. Box 9213, Ste. 3001-3004, Hui Bao, Long Commerce Center, Beijing, Capital, 10007-6, China
Phone: 86(10)67987599 Fax: 86(10)67987619
e-mail: info@e-cchina.com Web site: www.e-cchina.com
Contact: James Wei

Transco Technology Co., Ltd.
B406 Grand Pacific Garden Mansion, 8 Guanghua Road, Beijing 10026, China
Phone: (8610) 6581 9699 Fax: (8610) 6581 9799
e-mail: info@Transco.com.cn Web site: www.Transco.com.cn
Contact: Vic Dickson
Founded 1998. Member: ATA, LISA. Accepts resumes. Translates only Chinese into and from English, both traditional and simplified. Mostly technical, software. Has more than 100 in-house staff.

Denmark

eb Graphics Design, Boserupvej 13, Farum, Denmark
Phone: 45 44 99 66 36 Fax: 45 44 99 66 37
e-mail: dkgraph01@attglobal.net
Website: www.users.cybercity.dl/~dko1501/
Contact: Boris Christens-en
Founded 1993. Member: ATA. Accepts resumes. Translates Swedish into Danish and English into Danish. Mostly IT, Technical, marketing and automotive. Has 6 translators.

Network Communications, A/S, P.O. Box 2034, Hobrovej 134, Randers, DK-8900, Denmark
Phone: 45(86)407600 Fax: 45(86)424311
e-mail: network@netcomms.dk Website: www.netcomms.dk
Contact: Robin Brown

France

Medialocate, 31 Rue Tourat, Bordeaux, 33000, France
Phone: 33(5)56822828 Fax: 3(3)556811818
e-mail: info@medialocate.com Website: www.medialocate.com
Contact: James A. Anderson

Germany

Fachübersetzungen C.P. MacKusick, Im Eck 4, Niederweiler,
D-55491, Germany Phone: 0049(6543)6062 Fax: 0049(6543)6286
e-mail: cpm@mackusick.de Website: www.mackusick.de
Contact: Christine P. MacKusick

Localize-IT Communication Services GmbH, Lohenstr. 8, 82166 Graefelfin-g, Germany Phone: 49(0)898786-0
Fax: 49(0)898786-33
e-mail: info@Localize-it.de Website: www.Localize-it.com
Contact: Rolf W. Blum
Founded 1999. Member: ATA, BDUE, STC, LISA. Accepts resumes. Translates English and German into all European language. Mostly IT, telecom and software. Maintains pool of 120 translators.

SAP AG, Postfach 1461, Walldorf, 69185, Germany
Phone: 49(6227)741830 Fax: 49(6227)744260
e-mail: hanno.holzheuser@sap.com Website: www.sap.com
Contact: Hanno Holzheuser

Scholl Translations International, Nuen Morgen 1, Remseck, 71686, Germany Phone: 49(7146)821067 Fax: 49(7146)3780
e-mail: scholl@aol.com Website: www.scholl-translations.com
Contact: U.L. Scholl-Ogholoh

SciTech Communications GmbH, Im Weiher 12, Heidelberg, D-69121, Germany Phone: 49(6221)905040 Fax: 49(6221)161935
e-mail: info@scitech.de Website: www.scitech.de
Contact: Brigit Stäudel
Founded 1994. Member: ATA. Accepts resumes. Main language pairs are German into English, French, Italian, Spanish and Portuguese. Mostly medical, technical and scientific. Pool of about 50 translators.

Siemens Language Services, Steinstr. 14, Velbert, 42553, Germany
Phone: 49(5251)8-11558 Fax: 49(5251)2053-504002
e-mail: ilona.volmer@mch20.sbs.de
Website: www.siemens.com/language_services
Contact: Helen Volmer

XTRA Translation Services, Calwer Str. 7, Boeblingen, 71034, Germany Phone: 49(7031)2136653 Fax: 49(7031)2136677
e-mail: stefan.lampert@xtra-services.com
Website: www.xtra-services.com
Contact: Stefan Lampert

Hong Kong

Mutilingual Translation Services, 8/F, Sing Hing Commercial Bldg., 139, Wing Lok St., Sheung Wan, Hong Kong, Hong Kong
Phone: (852)25819099 Fax: (852)25819066
e-mail: syhkmts@netvigator.com Website: www.multilingual.com.hk
Contact: Stephen Viu

India

International Academy of Foreign Languages, S-351 Greater Kailash Part-I, New Delhi, 110048, India
Phone: 91(11)6475810 Fax: 91(11)6235557
e-mail: iafl@nde.vsnl.net.in Website: www.iafltrans.com
Contact: Narender Parwanda
Founded 1989. Member: ATA. Does not accept resumes. Translates mainly Hindi, Urdu, Bengali, Gujarati, Arabic, Punjabi, Tamil, Nepali, Marathi, Dari, Persian and Pushtu. Mostly business, telecommunications, medical, legal and IT. Has a pool of 800 translators.

Ireland

Alchemy Software Development, Harcourt Business Centre, Block 2, Harcourt Street, Dublin 2, Ireland
Phone: +353-1-708-2800 e-mail : info@AlchemySoftware.ie
Web Site : www.AlchemySoftware.ie
Contact: Anthony G. Powch

Italy

Interlanguage S.R.L., Strada Scaglia Est 134, Modena, MO 41100, Italy
Phone: 39(059)344720 Fax: 39(059)344300
e-mail: info@interlanguage.it Website: www.interlanguage.it
Contact: Susanna Ferretti
Founded 1986. Member: ATA. Accepts resumes. Main languages are Italian, German, French, Spanish, Portuguese and Dutch. Mostly mechanics, electronics, textiles, ads, marketing, legal, finance, economics and tourism. Has pool of about 500 translators.

Lexikon, Via Carlo Cattaneo 21, Napoli, 80128, Italy
Phone: 39(081)5586768 Fax: 39(081)5586737
e-mail: info@lexikon.it Website: www.lexikon.it

Lexis Srl, Via Dario Campana 56/A, Rimini, 47900, Italy
Phone: (39)0541790190 Fax: (39)0541790685
e-mail: lexis.rimini@lexisitalia.com Website: www.lexisitalia.com
Contact: Roberto Ganzerli

Japan

Techlingua, Inc., 10-3-206, Denen-Chofu 1-chome, Tokyo, 145-0071, Ota-ku, Japan Phone: 81(3)37225211
Fax: 81(3)37215223
Contact: Atsushi Tomii

TranNet.K.K., 26-40 Shinjuku, 7–Chome, Shinjuku-Ku, Tokyo, 160-0022, Japan Phone: 81(3)52875881
Fax: 81(3)52875882
e-mail: hamaji@trannet.co.jp Website: www.alc.co.jp/trannet
Contact: Michio Hamaji

Malta

International Translation Agency LTD Malta, "Ajirce"/5, Testaferrata St., Msida, MSD 02, Malta
Phone: (356)322404 Fax: (356)319008
e-mail: corporate@translatedotcom.com
Website: www.translatedotcom.com
Contact: Rachid Titouah

Mexico

Comunicar Servicios Profesionales de Traducción & Interpretación, Joaquin Romo 109-103, Colonia Miguel Hidalgo, Mexico D.F. 14260, Mexico Phone: (5)6060389 Fax: (5)6650268
e-mail: comunica@avantel.net Contact: Maria Elena Cruces Ramirez

Netherlands

Alchemy Software Development, Harcourt Business Center, Bloch 2 Harcourt St., Dublin 2, The Netherlands Phone: (353)170-8280
Fax: (353)170-8280 e-mail: tonyod@alchemysoftware.com
Website: www.alchemysoftware.com Contact: Anthony G. Powch

WYO Translation Agency, P.O. Box 292, Emmasingel 51, Weert, 6001BA, The Netherlands
Phone: 31(495)537334 Fax: 31(495)543219
Website: www.wyo.nl

Peru

INTERCONTACT - PERU, Gral. Borgoño 518, Lima, 18 Peru
Phone: 51(1)4474157 Fax: (51(1)4474157
e-mail: interc6@attglobal.net Website: www.intercontactperu.com
Contact: Claudia Delgado Nachtigall

Portugal

Cial Lingua Service International, Avenue República, 41-8°, Lisbon, 1050 Portugal Phone: 351(1)7937610
Fax: 351(1)7960783
e-mail: cialis@mail.telepac.pt

Russia

Delos O.O.O., 10-14, Nagatinskaya Naberezhnaya, Moscow, 115533 Russia Phone: 007-0959552530 Fax: 007-0959552530
e-mail: russianize@delos.ru Website: www.delos.ru
Contact: Anna Evgrafova
Founded 1990. Member: ATA, LISA. Accepts resumes. Main languages are Russian, German, French, Spanish, Japanese, Portuguese and Chinese. Mostly petroleum engineering, electric engineering, telecom-munications and IT. Has a pool of 200 translators.

Singapore

Getlocalized.Com, 46a & 48a Bukit Pasoh Rd., Singapore, 089858 Singapore Phone: (65)2200633 Fax: (65)3237097
e-mail: deepali@teamworks.com.sg Website: www.getlocalized.com
Contact: Deepali Chaturvedi

Spain

Hermes Traducciones Y Servicios Lingüísticos, Ronda de Poniente,
6-2° E, Centro Empresarial "Euronova", Tres Cantos, Madrid 28760, Spain Phone: 34(91)8036294 Fax: 34(91)8035300
e-mail: hermestr@hermes.com Website: www.hermestrans.com
Contact: Juan José Arevalillo
Founded 1991. Member: ATA, LISA, ACT. Accepts resumes. Translates mainly Spanish from and into other European languages. Mostly computing, localization, legal, medical, finance and business. Has 20 in-house translators and 25 off-site.

Interglossa, S.C.P., C/Rosselló 34, 2°-2°, Barcelona, 08029, Spain
Phone: 34(93)5396910 Fax: 34(93)5396911
e-mail: interglossa@interglossa.com Website: www.interglossa.com
Contact: Sonia Carrera Martinez Member: ATA

Translationary, Corsega 281, 08008, Barcelona Phone 34 932176822,
Web site: www.translationary.com
Contact: Marcos von Haeften

Switzerland

CLS Corporate Language Services AG, Dianastrasse 6, P.O. Box 1003, Zurich, 8039 Switzerland Phone: 41(1)2066850 Fax: 41(1)2066890
e-mail: cls@cls.ch Website: www.cls.ch
Contact: Doris Marty-Albisser

Taiwan

American Translation Link (AT-Link), 8F-12, 2 Fuhsing North Rd., Taipei, 104 Taiwan
Phone: 886(2)27763577 Fax: (419)793-1279
e-mail: mwang@american-translation.com
Website: www.american-translation.com Contact: Muriel Wang

United Arab Emirates

Eman Translation Services, Beniyas Square, Shaikha Maryam Bldg., 1st Fl., Office 104, Dubai, 27750 United Arab Emirates
Phone: 00971(4)2247066 Fax: 00971(4)2247944
e-mail: ets@emirates.net.ae
Contact: Omer Bashier
Founded 1995. Accepts resumes. Translates mainly Arabic, French, German, Italian, Spanish, Farsi, Hindi, Chinese and Turkish. Mostly legal. Has pool of 10 and uses freelancers.

United Kingdom

Chambers Harrap Publishers Ltd., 7 Hopetoun Crescent, Edinburgh
EH7 4AY, United Kingdom
Phone: +44(0)131 556 5929 Fax: +44(0)131 556 5313
e-mail: admin@chambersharrap.co.uk
Website: www.chambersharrap.co.uk
Contact: Administrator
Founded 1994 (Chambers - 1890; Harrap - 1901). Member: ATA. Accepts resumes. Requires excellent language skills and experience of dictionary translation. Translates mainly English into French, Spanish, German, Italian and Portuguese. Mostly dictionaries and language guides, including slang, business and IT dictionaries. Uses 20-50 translators.

CTT Languages, Fieldgate, Coltstaple Lane, Horsham, RH13 7BB, United Kingdom Phone: 44(7000)005006
Fax: 44(7000)005007
e-mail: info@ctt-languages.co.uk Website: www.ctt-languages.co.-uk
Contact: Tristan Loraine

The Language Technology Centre, Ltd., 5-7 Kingston Hill, Kingston Upon Thames, Surrey, KT2 7PW, United Kingdom
Phone: 44(208)5492359 Fax: 44(208)9746994
e-mail: asirjean@langtech.co.uk Website: www.langtech.co.-uk
Contact: Adriane E.M. Rinsche

SDL International, Butler House, Market Street, Maidenhead, Berkshire SL6 8AA, United Kingdom
e-mail: cstoldt@sdlintl.com Website: www.sdlintl.com
Contact: Vendor Management Group

TECHNOTRANS, Ridgeside Farm, The Ridge, Corsham, Wiltshire SN13 9PL, United Kingdom, Phone: 44(0)1225 811 891 Fax: 44(0)12-25 811 416
e-mail: tmoa@ukgateway.net
Contact: Thomas Moser

APPENDIX F: COMPUTERS, TRANSLATION, AND THE INTERNET

Digital technology as it relates to translation and to almost anything else in today's world is in a constant state of change. In the nearly three years since this handbook was last revised, computers have become more dominant in the translation craft. More work is now being transmitted electronically by e-mail attachment rather than by fax. The Internet has become a routine tool for translators – from work search to word search. And a growing number of translators has become involved in the translation of such computer-based material as websites, a process now generally referred to as localization. All this has resulted in this newly revised and expanded chapter that looks at all the above developments and how they benefit or fail to benefit translators.

Once you have read this chapter you will be glad to find out that one thing has not changed, which should come as no surprise: translation continues to be a human, rather than a machine function. Computers are no closer now to replacing human translation than they were three years ago, or at any other time in the past. Notwithstanding the great benefits translators are deriving from the digital revolution, computers continue to be the tools, rather than the decision-makers of the translation process.

Computers and Related Equipment

As was mentioned before, only a few years ago a translator would use a pen or a typewriter to translate. In 1980, for example, the ultimate text-producing tool in the world was the IBM Selectric electric typewriter, which we thought we would proudly bequeath to our children and grandchildren. All of this changed forever with the birth of the word processor, and more specifically, the PC, or Personal Computer, which is getting better every year, as if it were a magical tool with boundless possibilities. Certainly the PC has changed the lives of translators, increasing their productivity and profitability three- and fourfold, and enabling them to receive and transmit work, look for answers to linguistic questions, communicate around the world, and take advantage of computer-assisted tools (see below) that in certain cases can save a great deal of time and effort.

Bare Minimum Hardware and the Wish List

The electronic age offers a great deal more than the word-processor. There is the fax, modem, optical scanner, e-mail, various types of translation software such as translation memory, and last but not least, the Internet. No doubt, it all seems quite overwhelming. Which is the best tool? How much should I spend? How much do I need?

Perhaps the best way to start is by asking two questions: What is the bare minimum a translator needs to get started, and, if money were no object, what would be my wish list of electronic equipment?

The first thing you need to get is a good, current personal computer. This is a must. The question is whether to get a desktop PC, or a portable notebook? The desktop PC is bigger,

heavier, has a larger, easier-to-read screen, and an easier-to-use keyboard. The notebook, on the other hand, only weighs a few pounds, can fit in your briefcase, can be used anywhere (it has a rechargeable battery that gives you several hours of use), and, like a toothbrush, becomes a truly personal item. This book is being revised on a three-year-old IBM Think-Pad notebook. Though thin (1.8 inches) and flimsy-looking, it is a real powerhouse. The base price was $1,699.00. For $99.98 I upgraded the memory from 32MB RAM (see glossary in the back of the book) to 64MB. For $225.98 I got a second battery for more cordless computing hours while traveling. For $237.98 I got a 3-year warrantee for parts and service. It has a 266MHz Intel Pentium II processor, 3.2GB of hard drive space (upgradable to 6.4GB), a 56K modem, a

3.5 inch disk drive, and a CD-ROM drive which at the moment hosts a disk containing the entire Encyclopedia Britannica. The color monitor, a 12-inch SVGA, has good, sharp resolution. It operates on Microsoft Windows® 98. I use it mainly for word-processing, for e-mail, to surf the net, occasionally for graphics, and sometimes to hear music. At the moment it has some 25 programs, including WordPerfect 6.0 for Windows, MS Word, Lotus Notes, as well as my Internet provider, to mention only a few.

To me as a translator and a writer, this notebook is the best of all possible worlds. On the other hand, many translators prefer a desktop PC for all sorts of reasons, not all of which are known to me. Since I have access to both, it's hard for me to judge. My suggestion to you is to check around and try both, and make your own decision. But again, a translator without a computer in this day and age is like a painter without brushes. One more thing: The standard computer for translators is the IBM or IBM-compatible. The Macintosh, favored by graphics people, is not near-ly as widely used in our circles, so we will concentrate here on the former.

The above configuration of my notebook computer, which one can also get in a desktop PC (except for the batteries), is quite close to what a translator would consider a "wish list" configuration. Keep in mind that there is a vast difference between stor-ing text and storing graphics on a hard drive. One hundred pages of text take less disk space that one or two pages of graphics. If your translation work entails extensive graphics, you need a bigger hard drive and a faster computer, and the cost can go up 50%-100%. Most translators, however, do not have such a need. Instead, they can store data on their hard drive, and when it gets too crowde-d, they transfer files to a disk.

The other item that's a must is a printer. Although you can deliver text on disk or by e-mail, there are many good reasons why you need a printer. First, there is still a widespread requirement for hard copy. Second, you may want to print drafts to check your translation, and third, you need a printer for correspondence, for printing your resume, and certainly for printing an invoice. We recommend the Hewlett Packard family of printers, particularly the LaserJets. Depending on the volume, you can start with a smaller model and go up to the HP 4+ or a newer model, which is a bit faster and better.

The next item that's a must is the facsimile machine, popularly known as the fax. These days information travels very fast, and a sure way to get translation assignments is via the fax. Once you have translated a document, the fastest way to send it back is as an e-mail attachment. In other words, the basic translator tools today are the computer, the printer, the fax and the modem.

Another item you may want to look into is the optical scanner, starting with a recent model of the HP ScanJet family. Be sure to get it with the OCR, or Optical Character Reader. This optional piece of equipment provides two valuable functions—you can scan text from a page into a disk, which can be a great timesaving shortcut, and you can scan graphic-s, such as tables and photographs, into your text file.

What about the cost? You can easily spend around $5,000 on the above four items, if you go directly to the top of the line. But you don't have to start at the top of the line. You can start a couple of notches below the top, and still be able to give yourself a decent start.

You could, if you really want to watch your pennies, take advantage of the fact that compute-rs are chang-ing so fast, and look in the paper for a year or two old second-hand computer, but not older. You should be able to find a good one for $400-$800. You can do the same with the printer, by getting a second-hand one for about $100, or a new one for about $300-$400. A good modem (at least a 28.8 or higher) can be found for around $100, although you could spend $300-$400. A plain paper fax machine (thermal paper is out) can be found for around $200-$400, or you could start out by using a local service, like Kinko's. In short, you could get started with the basic equipment for as little as $600-$700 if you had to, and go on from there. As you prosper, you can start upgrading.

Other Useful Kinds of Software

You can use your word-processing software for doing record-keep-ing and invoic-ing as well, but if your operati-on becomes more involved you may want to consider a database program, such as Excel, for record-keep-ing (and also for other database-s, such as glossaries), and some basic invoicing program, which only runs $20-$30.

To protect your computer against computer viruses—the dreaded new computer age affliction—you may want to get a program such as Norton AntiVirus. To be able to retrieve lost text, you need the Norton Utilities program. Both are excellent safety measures. Another safety measure is the constant backing up of all your files. This can be done on the hard drive or on an external medium such as floppy disk or CD disk. Computers are great when they work, but when they fail they can cause you serious losses, which you cannot afford.

For word-processing, translators uses mainly Microsoft Word, but also WordPerfect, the old favorite.

Foreign Language Software

English-language word-processing is the most advanced in the world. Great strides have been made in programs for other languages, but there is still a long way to go. Some languages, such as Russian, Chinese, Japanese, Arabic and Hebrew are simply too complex in either their script or their noun and verb forms or both to lend themselves to, say, spell-check as readily as English. So don't expect to "cruise" as easily in those languages as you do in English. However, if you translate text from English into almost any other language, you are expected these days to deliver the text on electronic media, which means getting involved in word-processing in that language (see Appendix 3, Foreign Language Software Sources).

As a translator from English into a foreign language, you need software that accommodates not only your target-language, but also the source-language, namely, English. Sometimes you'll need to include English text in the body of your Chinese or Arabic text. Also, you have to make sure your foreign language software is compatible with English-language software, so that it can interface with your client's equipment. Therefore, we are focusing here on American-made foreign language software, rather than foreign-made.

Foreign Language Software Sources

In recent years the leading source for European languages software was WordPerfect. Its language modules covered a large variety of languages, and there were frequent upgrades. The WordPerfect foreign language modules are still used by many translators, but the trend is shifting to the localized products of Microsoft. These products use the Unicode system which makes Microsoft's English word processing compatible with its foreign versions, enabling the user to use English and foreign word processing in the same document with relative ease. Microso-ft's foreign language word processing is now part of its package of English word processing, which eliminates the high cost only a few years ago of buying foreign language modules separately.

Machine Translation Vs. Human Translation

Machine Translation, or MT, is the term used to describe trans-lation performed by a computer software program, as an alter-native to human translation (HT), performed by a human translator. Machine Translation belongs in the area of artificial intelligence. Artificial intelligence is the branch of computer science that deals with using computers to simulate human thinking. It looks to create programs that can solve problems *creatively,* rather than merely respond to commands. In other words, to operate just like the human brain.

In the late 1950s people in organizations such as the U.S. Air Force believed that computers would soon be pro-grammed to accept human language input and translate it into English or into any other language. During the following 50 years, millions of dollars were spent by the Air Force and by other U.S. Government bodies and by big business in the hope of having computers take over the human function of translation. So far, the results have been quite limited, for two main reasons: (a) while computers have a seemingly unlimited capacity for processing data, they are a long way from having the capacity to think creatively like human beings; and (b) human language is not merely a collection of signs and symbols that can be easily programmed, manipulated, and computerized. This is true of human language used not only in poetry and philosophy, but also in technical subjects, in which language expresses thought processes far more complicated than merely "one plus one equals two."

Many people today maintain that computers will soon replace translators. Most of those people are not translation experts. In fact, hardly any of them are. Some point to the fact that computers have already replaced typists and secretaries and reduced the work force of many companies. All this reminds me of a remark by George Bernard Shaw, according to which a monkey could write *Hamlet* if it managed to hit all the right keys on the typewriter. Very few translators can lay claim to creating literary works on the order of *Hamlet*, but lumping translators with typists and office support staff is missing the whole point of translation. As long as language continues to communicate more than the immediate literal meaning of words, as long as there are shades of meaning that keep changing all the time, as long as people have to make value judgments about the meaning and intent of a text, one will continue to need human translators to get the job done.

In recent years several companies in the U.S. and around the world have produced software designed to translate from one language into another. This software varies from a very basic word finder for the tourist, to complex programs for translating technical and scientific data. Some of the latter range in price from as low as $250 to as high as $250,000, and have been sold to governments and international organizations. One important lesson, however, has been learned by both the makers of those programs and their consumers. Machine translation does not replace human translation. At best, the former can achieve around 60 percent accuracy, when the goal is as close to 100 percent as possible. Consequently, the expectations regarding MT have been modified, and it is now recognized that to achieve full accuracy such translations must be post-edited by a human translator.

All of the above notwithstanding, Machine Translation does have its uses. Certain limited language environments do allow for machine translation. One example is the Canadian weather bureau, which transmits weather reports in both English and French. The number of words involved in the daily weather report is very limited, and can be easily programmed into a computer for translation from one language into another. Another example is official forms which contain simple basic questions. An organization such as NATO can have forms put out in ten different languages and set up a program to automatically translate each form into those languages. A third example is the Caterpillar company in Illinois that sells its agricultural machinery around the world and maintains operation manuals in various languages. This company has invested millions in its in-house translation software which allows it to update and modify its technical literature in a cost-effective way. In all of the above examples the software was customized for one particular work environment where repetition is the common denominator.

Another use of machine translation is for processing a large body of a foreign language text to find out what the gist of it is, rather than to achieve a fully accurate translation, in order to decide whether on not to select parts of it for accurate (viz., human) translation. This process is only partially reliable.

Translators for the most part are not fond of post-editing machine translation. Often the pay is not adequate, and the work can be harder and more time consuming than translating directly from the original. This certainly applies to freelancer-s, but also to in-house translators who are on a salary.

At the present time the ones who benefit the most from machine translation are not those who buy such services, but rather the companies who develop and manufacture the software and the translators who work on those machine translation projects either as developers of the programs or editors of the machine-translated text. In other words, machine translation has not made translation cheaper for the average consumer of translation services, and therefore does not pose a real threat to translators.

Translation Memory

One computer tool commonly used today in the area of computer-assisted-translation (CAT) is called translation memory. Here sentences and other parts of translated texts are stored and can be used again when the same or similar translation is required. This tool is best applied to major projects, such as automotive operation manuals that are updated or modified on a regular basis, whereby most of the text remains the same while certain sentences and/or paragraphs are changed.

While this tool has its practical uses, it has been hyped beyond its actual application and has been promoted as a universal aid for translators in nearly all types of translation. Here one should be careful not to invest in this kind of software unless one has a clear ongoing use for it. This applies both to individual translators and to translation companies, since in both cases only a small percentage of the work done is of the above-described kind.

Furthermore, as in the case of machine translation and localization (see below), here too the tools are far from able to accomplish their stated goals, and are still being improved upon. Translation memory is certainly not human memory, and while it does work part of the time, human memory remains irreplaceable.

Localization

Localization is defined as

the process of creating or adapting a product to a specific locale, i.e., to the language, cultural context, conventions and market requirements of a specific target market. With a properly localized product a user can interact with this product using his or her own language and cultural conventions. It also means that all user-visible text strings and all user documentation (printed and electronic) use the language and cultural conventions of the user. Finally, the properly localized product meets all regulatory and other requirements of the user's country or region.

Clearly, this definition takes us out of the strict realm of translation into the domain of computer programming and international business. It alludes primarily to the localization of websites (as well as software) in other languages and cultures. Two other terms often used in conjunction with localization are internationalization and globalization. The first refers to the preparation of computer text to be used in localization in such a way that it meets the requirements of the target language and culture. Globalization generally refers to both processes together, namely, internationalization and localization.

With the fast proliferation of websites and the rapid growth of many kinds of computer software, a growing number of translation companies have ventured into the field of localization, and more than a few computer hardware- and software-producing companies (including giants like IBM and Microsof-t) have turned to the field of translation as a function of localization.

Thus, localization has been presenting translators with new opportunities and some difficult questions. On the one hand, there is a huge volume of text these days in website and software development that necessitates multilingual translation. On the other hand, for a translator to become involved in localization a specialized knowledge of computers is usually required, since website and software localization requires not only language translation but also adaptation of computer commands to the requirements of the target language. Hence, a translator may want to carefully consider how far he or she may wish to become involved in technical computer work that requires skills and training beyond the scope of actual translation work.

It should be mentioned that localization in the sense of translating a text and adapting it to another culture existed long before the computer age. Thus, for example, an American advertisement company in the pre-computer era that planned an international ad campaign for a company such as Coca Cola, had to study the cultures of other countries and customize the ads not only in terms of the target language, but also the cultural conventions, preferences, taboos etc. of the target country. Furthermore, the act of translating itself is in a sense a form of localization, since language does not exist separately from its culture.

Website and software localization has been around now for several years. It relies heavily not only on human translation but also on machine and machine-assisted translation tools, such as translation software and translation memory. Given the limitations of these tools, the often unrealistic expectations of the marketplace and the eagerness to process a large volume of material in record time, localization has been experiencing many problems, and the goals its pursuers have set for themselves are yet to be realized. One should keep all of this in mind before getting involved in this area of translation.

The Uses of the Internet for Translators

The translation field has become increasingly "wired," witnessing a proliferation of sites relating to professional activities, job-hunting, commercial products, and online reference mate-rials for translators. It is not an exaggeration to say that trans-lators do themselves (and potentially their clients) quite a disservice by ignoring the virtual translation community.

This section will introduce you to a selection of highly-regarded and well-maintained sites of use to translators. It assumes basic Web literacy. Of course, the Internet's ever-changing nature guarantees that certain websites will decline or disappear, while new ones will emerge by the time this book becomes available. So please be aware that you will probably have to do some updating of the following information on your own.

Where We Are on the Web

First, let us invite you to Schreiber Publishing's website. Our address: **schreiberpublishing.com**. Its objective is to provide reference literature, training materials, and useful information for translators.

Professional Activities

Translators wishing to learn about the state of their profession will benefit from a visit to the American Translators Association's
site, **www.atanet.org**. I discuss the ATA in greater detail in the chapter on Translators' Organizations; its Website includes information on the Association's formal activities such as conferences, publications and accreditation procedures. Even if you are not an ATA member, the site is useful because it can guide you to other translation-related sites.

Another major website of interest to translators is the site of the
international umbrella organization of the world's translator associations, the International Federation of Translators, found on **www.fit-ift.org**.

Aquarius, at **www.aquarius.net.** Claims to be "the world's language network." This commercial website lists translation agencies seeking translators and interpreters who, in turn, can post resumes to make their services known. It also links to other translation sites.

Networking with Other Translators

An extremely popular program on the Internet for translators has been the *Language Forum* of Compuserve, popularly known as FLEFO (Foreign Language Education Forum). You have to subscribe to this forum to be able to participate. You enter FLEFO in the keyword box and you find yourself in a virtual forum full of translators from around the world, who post messages and get answers to almost any question a translator may ask, from the cost of a dictionary, to the meaning of a word, to a social get-together. FLEFO can also link you to translation wesites.

Another online forum for translators is the newsgroup: subscribers post messages and receive answers (say, to a specialized translation query). A newsgroup for translators is sci.lang. translati-on.

Reference Materials

The Translator's Home Companion at **www.rahul.net/lai/ companion.html**, provides Links to glossaries, dictionaries, translation publications, and other translators.

The Human Languages Page at **www.june29.com/IDP** provides links to other translation-related sites, including some listed here.

Language Today, an online publication (also available in print; see Bibliography appendix) covering language technology, translation and interpreting: **logos.it/language_today**

Online Dictionary Services

Eurodicautom at **www.europa.eu.int/eurodicautom** is a very useful service, Eurodicautom handles European languages such as Spanish, Portuguese, French, German, Italian, English, Danish and Dutch. Includes many specialized dictionaries in non-technical, legal and technical areas. Terms are both translated and defined.

Online Dictionaries at **www.yourdictionary.com** covers a far greater range of languages than the above entry. It provides links to dictionaries and translations in many specializations.

Glossaries

There are many mediocre glossaries out there in cyberspace, so use caution with this type of online resource. Here are some better ones:

www.refer.fr/termisti/liste.htm

www.yahoo.com/references/dictionaries

U.S. Government agencies provide some very useful glossaries. For example:

www.epa.gov (for environmental terminology).

Finding Cultural Information

A quick way to access cultural and historical information on other countries is via The Electronic Embassy at
www.embassy.org. This service provides links to embassy and/or United Nations home pages of many nations. Most countries' pages include links to cultural resources such as online newspapers, libraries, and so on. You may have to do some digging to find the information you're after, but if it exists online, you will probably find it by starting here. Also the foreign ministries of various countries are a good source for all sorts of information related to that particular country.

Shopping for Dictionaries

A prime source for dictionaries is i.b.d. Ltd.'s website is:
www.ibdltd.com.

Another site under construction is: **www.schoenhofs.com**.

Major virtual bookstores also have good selections of dictionaries (see box). Examples:

www.barnesandnoble.com
www.amazon.com
www.borders.com

You may also wish to consult these websites of major dictionary publishers:

Elsevier at **www.elsevier.com**
Routledge at **www.routledge.com**
McGraw-Hill at **www.mcgraw-hill.com**

Machine Translation

If you want to experience some machine translation, try these two sites:
www.babelfish.com/Languages/English/EnglishMachine.sht-ml
http://babelfish.altavista.com

Work Sources on the Internet

A European-based website that offers work sources and a newsletter with work information for translators is:
www.transtree.com
Three websites offer the capability to communicate with other freelancers and find out about the payment practices of translation agencies. For the most part they are free, but you have to join to be able to use them:

www.trwenterprises.com/payment_practices.htm
www.topica.com/lists/transpayment/
www.tcrlist.com/

Order Form

Fax orders: (301) 424-2336. Send this form.

Telephone orders: Call 1(800) 822-3213 x28 (in Maryland: 301-424-7737 x28)

E-mail orders: books@schreiberpublishing.com Mail orders to: Schreiber Publishing, 51 Monroe St., Suite 101, Rockville MD 20850 USA Website orders: www.schreiberpublishing.com or: **www.fortranslators.com**

Please send the following books, programs, and/or a free catalogue. I under stand that I may return any of them for a full refund, for any reason, no questions asked:

❑ **The Translator's Handbook 5th Revised Edition** - The Leading Guide for Translators - $25.95

❑ **The Translator's Self-Training Program** - includes work sources CD (circle the language/s of your choice):

Spanish French German Japanese Chinese Italian Portuguese Russian Arabic Hebrew - $69.00

❑ **The Translator's Self-Training Spanish Medical** - $69.00

❑ **The Translator's Self-Training Spanish Legal** - $69.00

❑ **The Translator's Self-Training German Patents** - $69.00

❑ **The Translator's Self-Training Japanese Patents** - $69.00

❑ **Multicultural Spanish Dictionary** - How Spanish Differs from Country to Country - $24.95

❑ **21st Century American English Compendium 3rd Rev. Ed.** - The "Odds and Ends" of American English - $29.95

❑ **Spanish Business Dictionary** - Multicultural Business Spanish - $24.95

❑ **French Business Dictionary** - Business terms from France and Canada - $24.95

❑ **Russian Business Dictionary** - $24.95

❑ **Japanese Business Dictionary** - $24.95

❑ **Chinese Business Dictionary** - $24.95

❑ **German Business Dictionary** - $24.95

❑ **Global Business Dictionary** - $33.95

❑ **Spanish Chemical and Pharmaceutical Glossary** - $29.95

❑ **Dictionary of Medicine French-English with English-French Glossary** - $179.50

❑ **Translator Work Sources** (CD and mailing label lists) - $69.00

❑ **Glosario internacional para el traductor 4th Revised Edition** - $44.35

❑ **Diccionario de la lengua española** - 2 volume economy edition - $84.00

❑ **Diccionario de la lengua española** - one volume large print - $195.00

❑ **La traducción del inglés al catellano** - $21.25

❑ **Diccionario bilingüe de mejicanismos** - $29.95

❑ **Spanish False Cognates** - $29.95

Sales tax: Please add 5% sales tax in Maryland

Shipping (est.): $4.50 for the first book and $2.00 for each additional product

International: contact us for shipping rates Order total: ______________

Name: ____

Address: ____ City: ____________________

State: _______ Zip: ____________ Telephone: _____________________ e-mail: ____________________

Payment: ❑ Cheque ❑ Credit card: ❑ Visa ❑ MasterCard

Card number: ____ Name on card: ________________________

Exp. Date:_____ /_____